HUTCHINSON

Dictionary of

Geography

Titles in this series

Dictionary of Abbreviations
Dictionary of Biology
Dictionary of Chemistry
Dictionary of Classical Music
Dictionary of Computing and Multimedia
Guide to Countries of the World
Dictionary of Geography
Dictionary of Mathematics
Dictionary of Physics
Dictionary of Science
Dictionary of World War I
Dictionary of World War II
Dictionary of 20th Century World History
Dictionary of Battles
On This Day
Chronology of World Events
Family Encyclopedia

HUTCHINSON

Dictionary of Geography

BROCKHAMPTON PRESS
LONDON

Helicon Publishing Ltd
42 Hythe Bridge Street
Oxford OX1 2EP

Printed and bound in Great Britain by
Mackays of Chatham Plc,
Chatham, Kent

This edition published 1997 by
Brockhampton Press Ltd
20 Bloomsbury Street
London WC1B 3QA
(*a member of the Hodder Headline PLC Group*)

ISBN 1-86019-572-5

British Cataloguing in Publication Data

A catalogue record for this book is available
from the British Library

Editorial director
Michael Upshall

Consultant editors
Simon Ross
Michael Thum

Project editor
Sara Jenkins-Jones

Text editor
Catherine Thompson

Art editor
Terence Caven

Additional page make-up
Helen Bird

Production
Tony Ballsdon

A

ablation the loss of snow and ice from a ◊glacier by melting and evaporation. It is the opposite of ◊accumulation. Ablation is most significant near the snout, or foot, of a glacier and round its edges, since conditions tend to be warmer at lower altitudes. The rate of ablation also varies according to the time of year, being greatest during the summer. If total ablation exceeds total accumulation for a particular glacier, then the glacier will retreat; if total accumulation exceeds total ablation, the glacier will advance.

abrasion the effect of ◊corrasion, a type of erosion in which rock fragments, carried by rivers, wind, ice, or sea, scrape and grind away a surface. ◊Striations, or grooves, on rock surfaces are a common abrasion, caused by the scratching of rock debris embedded in glacier ice.

accessibility the ease with which a place may be reached. An area with high accessibility will generally have a well-developed transport network and be located centrally or at least at a ◊route centre. Many economic activities, such as retailing, commerce, and industry, require high accessibility for their customers and raw materials.

Accessibility can be measured by an accessibility index, such as the ***Shimbel index***. The Shimbel index of each place in a given transport network may be worked out by producing a simplified ◊topological map of the network and then constructing a table showing the number of links necessary to get from one destination to another. The place with the fewest links has the lowest Shimbel index and, therefore, the highest accessibility. This method ignores all other factors, such as population density, distance, quality of link, and traffic flow. See also ◊connectivity.

accumulation the addition of snow and ice to a ◊glacier. It is the opposite of ◊ablation. Snow is added through snowfall and avalanches, and is gradually compressed to form ice as the glacier progresses.

accessibility

topological map of the road network linking towns and cities of W England and SE Wales

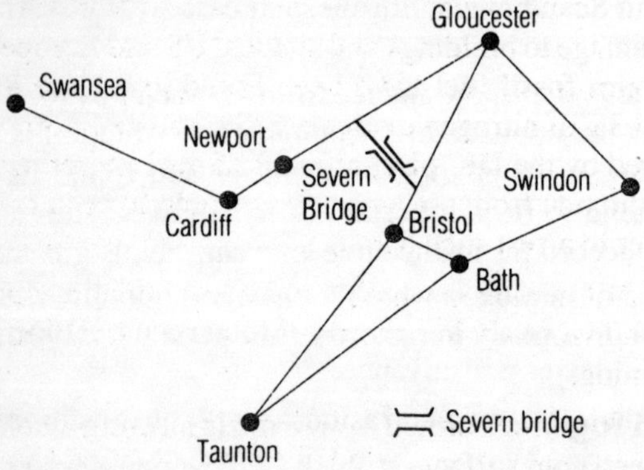

accessibility index table

	Bath	Bris	Card	Glou	Newp	Swan	Swin	Taun	Total
Bath	–	1	3	2	2	4	1	1	14
Bris	1	–	2	1	1	3	1	1	10
Card	3	2	–	2	1	1	3	3	15
Glou	2	1	2	–	1	3	1	2	12
Newp	2	1	1	1	–	2	2	2	11
Swan	4	3	1	3	2	–	4	4	21
Swin	1	1	3	1	2	4	–	2	14
Taun	1	1	3	2	2	4	2	–	15

Although accumulation occurs at all parts of a glacier, it is most significant at higher altitudes near the glacier's start, where temperatures are lower.

acid rain acidic rainfall, thought to be caused principally by the release into the atmosphere of sulphur dioxide (SO_2) and oxides of nitrogen.

Sulphur dioxide is formed from the burning of fossil fuels, such as coal, that contain high quantities of sulphur; nitrogen oxides are contributed from various industrial activities and from car exhaust fumes.

Acid rain is linked with damage to and the death of forests and lake organisms in Scandinavia, Europe, and eastern North America. It also results in damage to buildings and statues. US and European power stations that burn fossil fuel have been found to release 8 g of sulphur dioxide and 3 g of nitrogen oxides per kilowatt-hour. According to figures prepared by the UK's Department of the Environment, emissions of sulphur dioxide from power stations would have to be decreased by 81% in order to arrest such damage.

adit method of mining in which the mineral-bearing rock is excavated by digging horizontally into the side of a valley. It is used, for example, in ◊coal mining.

advection fog ◊fog formed by warm air meeting a colder current or flowing over a cold surface.

afforestation planting of trees in areas that have not previously held forests (***reafforestation*** is the planting of trees in previously forested areas). Trees may be planted (1) to provide timber and wood pulp; (2) to provide firewood; (3) to bind soil together and prevent soil erosion; and (4) to act as windbreaks.

Between 1945 and 1980 the Forestry Commission doubled the forest area in the UK – planting, for example, Kielder Forest in Cumbria.

age–sex graph another name for a ◊population pyramid, a graph demonstrating the age and sex structure of a population.

agglomeration the clustering of activities or people at specific points or areas; for example, at a ◊route centre. Firms or individuals that cluster together can often share facilities and services, resulting in lower costs (due to ◊economy of scale).

agribusiness ◊commercial farming on an industrial scale, often financed by companies whose main interests lie outside agriculture; for example, multinational corporations. Agribusiness farms are mechanized, large in size, highly structured, reliant on chemicals, and may be described as 'food factories'.

agriculture the practice of farming, including the cultivation of the soil (for raising crops) and the raising of domesticated animals. Crops are grown for human nourishment, animal fodder, or commodities such as cotton, and animals are raised for wool, milk, leather, dung (for fuel), or meat. The units for managing agricultural production vary from small holdings and individually owned farms to corporate-run farms (agribusinesses) and ◊collective farms run by entire communities.

In the 20th century, the increased demands made upon agriculture by a growing world population have led to greater land clearance and the intensification of farming methods. After World War II there was an explosive growth in the use of agricultural chemicals (agrochemicals), such as herbicides, insecticides, fungicides, and artificial, non-organic fertilizers. In the 1950s and 1960s ◊high-yield varieties of crops such as maize and rice were developed, which increased food production in the developing world (the ◊green revolution) but increased reliance on agrochemicals. At the same time the industrialized countries developed factory farming, in which poultry, pigs, and cattle are reared indoors under close confinement and fed a strictly controlled high-protein diet.

From the 1970s concern has grown about the effects of some farming methods, leading to a movement towards sophisticated natural, or organic, methods without chemical sprays and fertilizers. Nitrates in fertilizers can leach away from the soil to pollute water supplies, and pesticides can pass through ◊food chains, accumulating in the diets of animals, including humans. Factory farming has lowered the cost of meat, but has also aroused controversy about its cruelty and about possible health hazards such as salmonella food poisoning. Land clearance and ◊deforestation have destroyed the natural habitats of many animal and plant species and have also led to ◊soil erosion in which the top, fertile layer of soil – no longer anchored by tree and shrub roots – is blown or washed away, leaving behind a barren desert, or 'dust bowl'.

Greater efficiency in agriculture, coupled with government subsidies for domestic production in the USA and the European Community (EC), has led to the accumulation of huge surpluses, nicknamed 'lakes' (wine, milk) and 'mountains' (butter, beef, grain). There is no simple solution to this problem as any large-scale dumping of excess stocks onto the market would lower prices, adversely affecting small farmers. Increasing concern about the starving and the cost of storage has led the USA and the EC to develop measures for limiting production – such as the ◊land set-aside scheme to reduce grain stocks, and the introduction of ◊quotas.

aid money or resources given or lent on favourable terms to developing countries. A distinction may be made betwen ***short-term aid*** (usually food and medicine), which is given to relieve conditions in emergencies such as famine, and ***long-term aid***, which is intended to promote economic development and improve the quality of life – for example, by funding irrigation, education, and communications programmes.

In the late 1980s official aid from the governments of richer nations amounted to \$45–\$60 billion annually, whereas voluntary organizations, such as Oxfam, received about \$2.4 billion a year. All industrialized United Nations member countries devote a proportion of their gross national product (GNP) to aid, ranging from 0.20% of GNP (Ireland) to 1.10% (Norway) (1988 figures). The UK development-aid budget in 1988 was 0.32% of GNP, with India and Kenya among the principal beneficiaries.

air mass large body of air with particular characteristics of temperature and humidity. An air mass forms when air rests over an area long enough to pick up the conditions of that area. For example, an air mass formed over the Sahara will be hot and dry. When an air mass moves to another area it affects the ◊weather of that area, but its own characteristics become modified in the process. For example, a Saharan air mass becomes cooler as it moves northwards.

The weather of the UK is affected by a number of air masses which, having different characteristics, bring different weather conditions. An Arctic air mass, for example, brings cold conditions, whereas a Saharan air mass brings hot conditions.

airport landing strip with a facility for the loading and unloading of passengers and freight from aircraft. Commercial airports normally

air mass

air masses affecting the UK

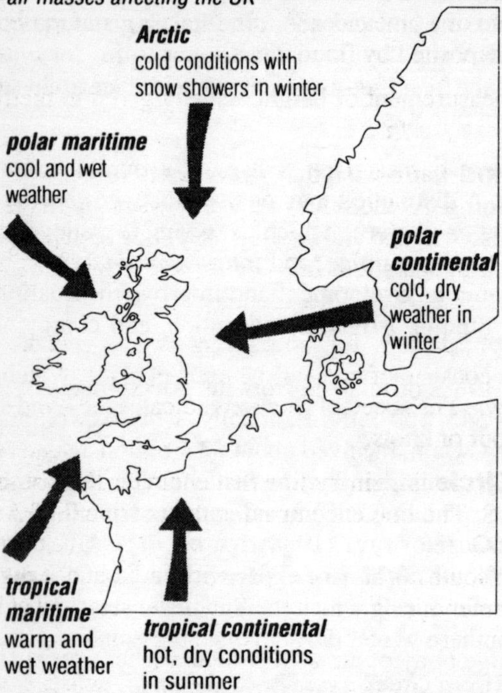

have a purpose-built terminal building, concrete or tarmac runways, and instrumentation such as radar for guiding and locating planes. A ◊stolport is a specialized type of airport.

Factors important in the location of airports are: (1) easy access via motorway and rail links; (2) nearness to urban areas for local demand; (3) flat land; and (4) suitable weather; for example, a large number of fog- and snow-free days.

alluvial fan roughly triangular depositional landform found at the base of slopes. An alluvial fan results when a sediment-laden river rapidly deposits its load of gravel and silt as its velocity is reduced on entering a plain.

alluvium fine silty material deposited by a river. It is deposited along the river channel where the water's velocity is low – for example, on the inside bend of a ◊meander. A ◊flood plain is composed of alluvium periodically deposited by floodwater.

altitude a measurement of height, usually given in metres above sea level.

anabatic wind warm wind that blows uphill in steep-sided valleys in the early morning. As the sides of a valley warm up in the morning the air above is also warmed and rises up the valley to give a gentle breeze. By contrast, a ◊katabatic wind is cool and blows down a valley at night.

anemometer device for measuring wind speed. A ***cup-type anemometer*** consists of cups at the ends of arms, which rotate when the wind blows. The speed of rotation indicates the wind speed in kilometres per hour or knots.

Antarctic Circle imaginary line that encircles the South Pole at latitude 66° 32' S. The line encompasses the continent of Antarctica and the Antarctic Ocean.

The region south of this line experiences at least one night during the southern summer during which the Sun never sets, and at least one day during the southern winter during which the Sun never rises.

anthracite (from Greek *anthrax*, 'coal') hard, dense, shiny variety of ◊coal, containing over 90% carbon and a low percentage of ash and impurities, which causes it to burn without flame, smoke, or smell.

Anthracite gives intense heat, but is slow-burning and slow to light; it is therefore unsuitable for use in open fires. Among the chief sources of anthracite coal are Pennsylvania in the USA; S Wales, UK; the Donbas, Ukraine; and Shanxi province, China.

anticline ◊fold in the rocks of the Earth's crust in which the layers or beds bulge upwards to form an arch. The opposite of an anticline is a ◊syncline.

anticyclone area of high atmospheric pressure caused by descending air, which becomes warm and dry. Winds radiate from a calm centre, taking a clockwise direction in the northern hemisphere and an anticlockwise direction in the southern hemisphere. Anticyclones are characterized by clear weather and the absence of rain and violent winds. In summer they bring hot, sunny days and in winter they bring fine, frosty spells, although fog and low cloud are not uncommon in the UK. ***Blocking anticyclones***, which prevent the normal air circulation of an area, can cause summer droughts and severe winters.

For example, the summer drought in Britain in 1976, and the severe winters of 1947 and 1963 were caused by blocking anticyclones.

appropriate technology or ***intermediate technology*** simple or small-scale machinery and tools that, because they are cheap and easy to produce and maintain, may be of most use in the developing world; for example, hand ploughs and simple looms. This equipment may be used to supplement local crafts and traditional skills to encourage small-scale industrialization.

Many countries suffer from poor ◊infrastructure and lack of capital but have the large supplies of labour needed for this level of technology. The use of appropriate technology was one of the recommendations of the ◊Brandt Commission.

aquifer any rock formation containing water. The rock of an aquifer must be porous and permeable (full of interconnected holes) so that it can absorb water. Aquifers supply ◊artesian wells, and are actively sought in arid areas as sources of drinking and irrigation water.

arable farming cultivation of crops, as opposed to the keeping of animals. Crops may be ◊cereals, vegetables, or plants for producing oils or cloth. Arable farming generally requires less attention than livestock farming. In a ◊mixed farming system, crops may therefore be found farther from the farm centre than animals.

In the UK a major arable farming area is East Anglia, where it is favoured by flat land, fertile well-drained soils and a warm and sunny climate. Many arable farms practise ◊crop rotation to maintain soil fertility.

arch natural curved archway created by the ◊coastal erosion of a headland. It is usually formed when the backs of two caves, on either side of the headland, are broken through. The roof of an arch eventually collapses to leave part of the headland isolated as a ◊stack. An example of an arch is Durdle Door in Dorset, England.

Arctic Circle imaginary line that encircles the North Pole at latitude 66° 32' N. Within this line there is at least one day in the summer during which the Sun never sets, and at least one day in the winter during which the Sun never rises.

arête sharp narrow ridge separating two ◊glacier troughs. The typical U-shaped cross sections of glacier troughs give arêtes very steep sides. Arêtes are common in glaciated mountain regions such as the Rockies, the Himalayas, and the Alps.

In the UK, arêtes are to be found in the Lake District and Snowdonia.

arid region region that is very dry, with an annual rainfall of less than 250 mm (by comparison, London has an average rainfall of 600 mm). Because rainfall is so low, there is little vegetation. Arid regions may have high rates of ◊evaporation, which can mean that more water is lost from the land in the form of vapour than is received by ◊precipitation. There are arid regions in Morocco, Pakistan, Australia, the USA, and elsewhere. Very arid regions are ◊deserts.

artesian well well that is supplied with water rising from an underground water-saturated rock layer (aquifer). The water rises from the aquifer under its own pressure. Such a well may be drilled into an aquifer that is confined by impermeable rocks both above and below. Much use is made of artesian wells in E Australia, where aquifers filled by water in the Great Dividing Range run beneath the arid surface of the Simpson Desert.

aspect the direction in which a slope faces. In the northern hemisphere a slope with a southerly aspect receives more sunshine than other slopes and is therefore better suited for growing crops that require many hours of sunshine in order to ripen successfully. Vineyards in northern Europe are usually situated on south-facing slopes.

assembly industry manufacture that involves putting together many prefabricated components to make a complete product; for example, a car or television set. The inputs for this type of industry are therefore outputs from others. Some assembly industries are surrounded by their suppliers; others use components from far afield.

assisted area region that is receiving some help from the central government, usually in the form of extra funding, as part of a regional policy. Most policies concentrate on identifying and then assisting 'backward' or 'problem' areas so that economic activity may be more equally shared within the country.

Examples in the UK include former heavy manufacturing centres such as Manchester and Clydeside. Some are designated ◊enterprise zones.

atmosphere the mixture of gases that surrounds the Earth; it is prevented from escaping by the pull of the Earth's gravity.

The lowest level of the atmosphere, the ***troposphere***, is heated by the Earth, which is warmed in turn by radiation from the Sun. Warm air cools as it rises in the troposphere, causing rain and most other weather phenomena. The upper levels of the atmosphere, particularly the ***ozone layer***, absorb almost all of the ultraviolet light radiated by the Sun, and prevent lethal amounts from reaching the Earth's surface.

atmospheric pressure the pressure at any point on the Earth's surface that is due to the weight of the column of air above it; it therefore decreases as altitude increases. Pressure is measured in millibars or kilopascals, using a ◊barometer. At sea level the average pressure is 1,013 millibars (101 kilopascals). Areas of relatively high pressure (◊anticyclones) and low pressure (◊depressions) can be identified.

attrition process by which particles of rock being transported by river, wind, or sea are rounded and gradually reduced in size by being struck against one another.

The rounding of particles is a good indication of how far they have been transported. This is particularly true for particles carried by rivers, which become more rounded as the distance downstream increases.

avalanche (from French *avaler* 'to swallow') fall of a mass of snow and ice down a steep slope. Avalanches occur because of the unstable nature of snow masses in mountain areas.

Changes of temperature, sudden sound, or earth-borne vibrations may trigger an avalanche, particularly on slopes of more than 35°. The snow compacts into ice as it moves, and rocks may be carried along, adding to the damage caused.

Avalanches are particularly hazardous in ski resort areas such as the French Alps. In 1991 a massive avalanche considerably altered the shape of Mount Cook in New Zealand.

B

backwash the retreat of a wave that has broken on a ◊beach. When a wave breaks, water rushes up the beach as ◊swash and is then drawn back towards the sea as backwash.

bar deposit of sand or silt formed in a river channel, or a long sandy ridge running parallel to a coastline. Coastal bars can extend across estuaries to form ***bay bars***.

barley cereal plant resembling wheat but more tolerant of cold and drought. Barley was one of the earliest cereals to be cultivated, and no other cereal can thrive in so wide a range of climatic conditions. Barley is no longer much used in bread-making, but it is used in soups and stews and as a starch. Its high-protein form finds wide use as animal feed, and its low-protein form is used in brewing and distilling alcoholic drinks.

barometer instrument that measures atmospheric pressure as an indication of weather (see ◊depression and ◊anticyclone). The barometers most commonly used are the ***mercury barometer*** and the ***aneroid barometer***.

basalt commonest extrusive (volcanic) ◊igneous rock, and the principal rock type on the ocean floor. It is commonly formed at ◊constructive margins of plates and is usually dark grey, but can also be green, brown, or black.

Basaltic lava tends to be runny and flows for great distances before solidifying. Successive eruptions of basalt have formed the great plateaux of Colorado and the Indian Deccan. In some places, such as Fingal's Cave in the Inner Hebrides of Scotland and the Giant's Causeway in Antrim, Northern Ireland, shrinkage during the solidification of the molten lava caused the formation of hexagonal columns.

base level level, or altitude, at which a river reaches the sea or a lake. The river erodes down to this level. If base level falls, ◊rejuvenation takes place.

basti or ***bustee*** Indian name for an area of makeshift housing; see ◊shanty town.

batholith large, irregular, deep-seated mass of ◊igneous rock, usually granite. The batholith forms by intrusion or upswelling of magma through the surrounding rock. Batholiths form the core of all major mountain ranges. In the UK, a batholith underlies SW England and is exposed in places to form areas of high ground.

batholith

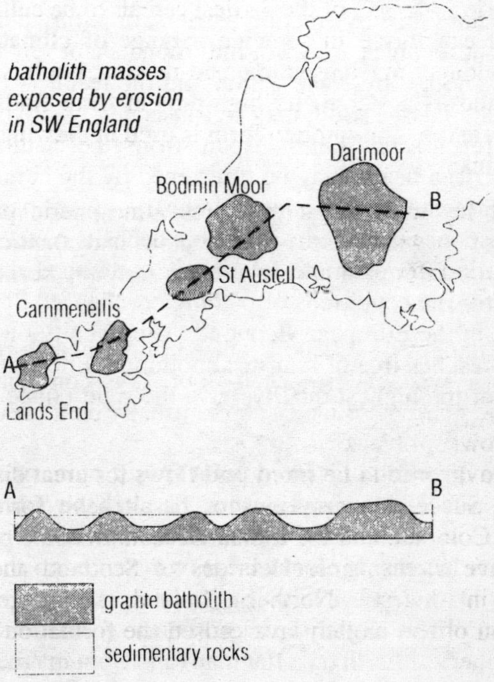

bauxite principal ore of aluminium, formed by the ◊chemical weathering of aluminium-bearing rocks in tropical climates. To produce aluminium the ore is processed into a white powder (alumina), which is then smelted by passing a large electric current through it. The chief producers of bauxite are Australia, Guinea, Jamaica, the Commonwealth of Independent States, Surinam, and Brazil.

beach strip of land bordering the sea, normally consisting of boulders and pebbles on exposed coasts or sand on sheltered coasts. It is usually defined by the high- and low-water marks; see ◊berm.

The material of the beach consists of a rocky debris eroded from exposed rocks and headlands. The material is transported to the beach, and along the beach, by waves that hit the coastline at an angle, resulting in a net movement of the material in one particular direction. This movement is known as ***longshore drift***. Attempts are often made to halt longshore drift by erecting barriers, or jetties, at right angles to the movement. Pebbles are worn into round shapes by being battered against one another by wave action and the result is called ***shingle***. The finer material, the ***sand***, may be subsequently moved about by the wind and form sand dunes. Apart from the natural process of longshore drift, a beach may be threatened by the commercial use of sand and aggregate, by the mineral industry – since particles of metal ore are often concentrated into workable deposits by the wave action – and by pollution (for example, by oil spilt or dumped at sea).

Concern for the conditions of bathing beaches led in the 1980s to a directive from the European Economic Community on water quality. In the UK, beaches free of industrial pollution, litter, and sewage, and with water of the highest quality, have the right (since 1988) to fly a blue flag.

bearing the direction of a fixed point from a point of observation on the Earth's surface, expressed as an angle from the north. Bearings are taken by a ◊compass and are measured in degrees (°), given as three-digit numbers increasing clockwise. For instance, north is 000°, south is 180°, and southwest is 225°.

True north differs slightly from grid north (because it is impossible to show a spherical Earth on a flat map) and from magnetic north.

bearing

bearings of the compass points

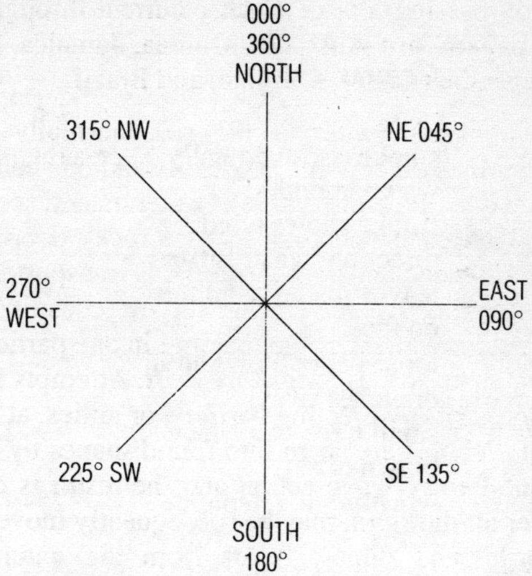

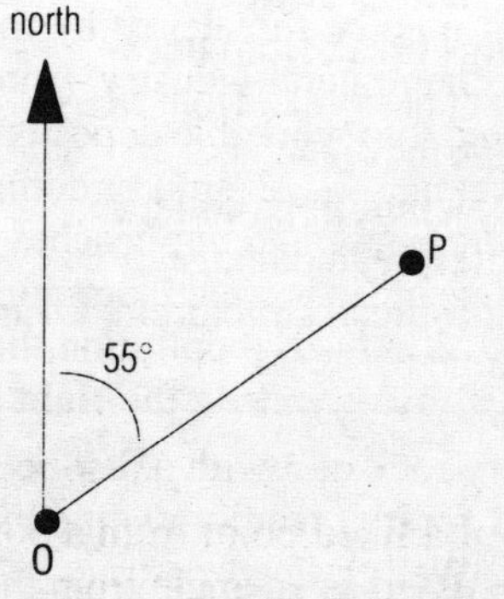

the bearing of P from O is 055°

The bearing of one point (P) from another (O) on a map can be found by measuring the angle between the line OP connecting O and P and the line pointing northwards from O.

Beaufort scale system of recording wind velocity, devised by Francis Beaufort 1806. It is a numerical scale ranging from 0 to 17, calm being indicated by 0 and a hurricane by 12–17.

Beaufort scale

number and description	*features*	*air speed* *mi per hr*	*m per sec*
0 calm	smoke rises vertically; water smooth	less than 1	less than 0.3
1 light air	smoke shows wind direction; water ruffled	1–3	0.3–1.5
2 slight breeze	leaves rustle; wind felt on face	8–12	3.4–5.4
3 gentle breeze	loose paper blows around	8–12	3.4–5.4
4 moderate breeze	branches sway	13–18	5.5–7.9
5 fresh breeze	small trees sway, leaves blown off	19–24	8.0–10.7
6 strong breeze	whistling in telephone wires; sea spray from waves	25–31	10.8–13.8
7 moderate gale	large trees sway	32–38	13.9–17.1
8 fresh gale	twigs break from trees	39–46	17.2–20.7
9 strong gale	branches break from trees	47–54	20.8–24.4
10 whole gale	trees uprooted, weak buildings collapse	55–63	24.5–28.4
11 storm	widespread damage	64–73	28.5–32.6
12 hurricane	widespread structural damage	above 73	above 32.7

bed single ◊sedimentary rock unit with a distinct set of physical characteristics or contained fossils, readily distinguishable from those of beds above and below. Well-defined partings called ***bedding planes*** separate successive beds or strata.

The depth of a bed can vary from a fraction of a centimetre to several metres, and can extend over any area.

bedload material rolled or bounced (by ◊saltation) along a river bed. The particles carried as bedload are much larger than those carried in suspension in the water. During a flood many heavy boulders may be moved in this way – such boulders can be seen lying on the river bed during times of normal flow.

bergschrund deep crevasse that may be found at the head of a ◊glacier.

berm on a beach, a ridge of sand or pebbles running parallel to the water's edge, formed by the action of the waves on beach material. Sand and pebbles are deposited at the farthest extent of ◊swash (advance of water) on a beach. Berms can also be formed well up a beach following a storm, when they are known as ***storm berms***.

beta index a mathematical measurement of the ◊connectivity of a transport network. If the network is represented as a simplified topological map, made up of nodes (junctions or places) and edges (links), the beta index may be calculated by dividing the number of edges by the number of nodes. If the number of nodes is n and the number of edges is e, then the beta index β is given by $\beta = e/n$

bid rent the price that a particular user is prepared to pay for a unit of land.

bid-rent theory assumption that land value and rent decrease as distance from the central business district increases. Shops and offices have greater need for central, accessible locations than other users (such as those requiring land for residential purposes) and can pay higher prices. They therefore tend to be located within the expensive central area.

The bid-rent theory may also be true for farming, with the most intensive use being made of the relatively expensive land on the outskirts of towns. Other factors, including ◊relief, communications, aspect, and land quality, may distort the relationship between price and location.

biofuel any solid, liquid, or gaseous fuel produced from organic (once living) matter, either directly from plants or indirectly from industrial, commercial, domestic, or agricultural wastes. There are three main ways for the development of biofuels: the burning of dry organic wastes (such as household refuse, industrial and agricultural wastes,

beta index

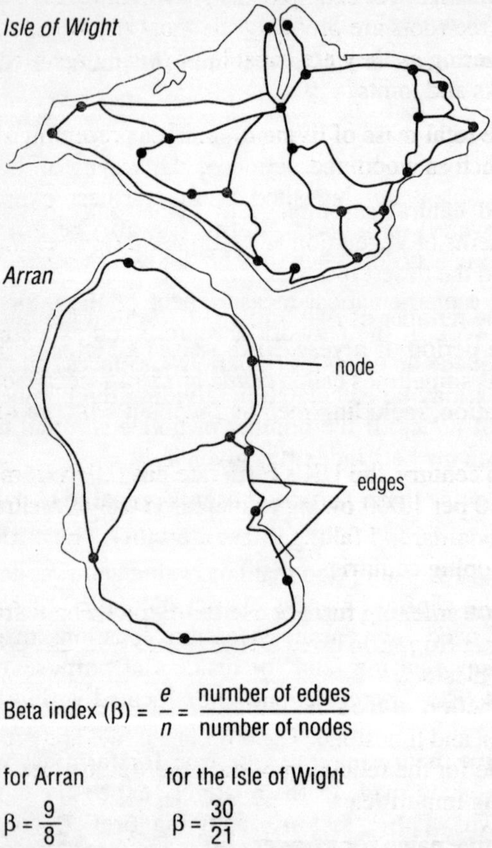

$$\text{Beta index } (\beta) = \frac{e}{n} = \frac{\text{number of edges}}{\text{number of nodes}}$$

for Arran	for the Isle of Wight
$\beta = \frac{9}{8}$	$\beta = \frac{30}{21}$
$\beta = 1.125$	$\beta = 1.43$

straw, wood, and peat); the fermentation of wet wastes (such as animal dung) in the absence of oxygen to produce biogas (containing up to 60% methane), or the fermentation of sugar cane or corn to produce alcohol; and energy forestry (producing fast-growing wood for fuel).

biological weathering form of ◊weathering caused by the activities of living organisms – for example, the growth of roots or the burrowing of animals. Tree roots are probably the most significant agents of biological weathering as they are capable of prising apart rocks by growing into cracks and joints.

biomass the total mass of living organisms present in a given area or ecosystem, such as woodland.

biome broad natural assemblage of plants and animals shaped by common patterns of vegetation and climate. Examples include the tundra biome and the desert biome.

birth rate the number of live births per 1,000 of the population of an area over the period of a year. Birth rate is a factor in ◊demographic transition. It is sometimes called ***crude birth rate*** because it takes in the whole population, including men and women who are too old to bear children.

In the 20th century, the UK's birth rate has fallen from 28 per 1,000 to less than 10 per 1,000 owing to increased use of contraception, better living standards, and falling infant mortality. The birth rate remains high in developing countries.

blast furnace smelting furnace used to extract metals from their ores, chiefly pig iron from iron orre. The temperature is raised by the injection of an air blast.

In the extraction of iron the ingredients of the furnace are iron ore, coke (carbon), and limestone. The coke is the fuel and provides the carbon monoxide for the reduction of the iron ore; the limestone acts as a flux, removing impurities.

bluff alternative name for a ◊river cliff.

bog type of wetland where decomposition is slowed down and dead plant matter accumulates as ◊peat. Bogs develop under conditions of low temperature, high acidity, low nutrient supply, stagnant water, and oxygen deficiency. Typical bog plants are sphagnum moss, rushes, and cotton grass. Large bogs are found in Ireland and northern Scotland.

boulder clay another name for ◊till, a type of glacial deposit.

braiding the subdivision of a river into several channels caused by deposition from islets in the channel. Braided channels are common in meltwater streams.

Brandt Commission international committee 1977–83 set up to study global development issues. The commission produced two reports, which stress that the countries of the wealthy, industrialized North and the poor South (or developing world) are dependent on one another. The reports suggested ways by which resources could be transferred to the Southern nations, together with measures that could be taken by the South to reduce poverty and increase food production.

The commission, officially named the Independent Commission on International Development Issues, had 18 members and was headed by West German chancellor Willy Brandt (1913–). The reports noted that measures taken in the past had met with limited success; this was also the fate of the commission's recommendations.

break-of-bulk point place where goods are transferred from one form of transport to another. This frequently involves the goods being repackaged into smaller quantities ready for individual users. Break of bulk often occurs at ports, where cargo carried on inland waterways is transferred to ocean-going vessels; Rotterdam is an example. Break-of-bulk points may be important industrial locations.

brownfield site site that has previously been developed; for example, a derelict area in the inner city. Before brownfield sites can be redeveloped, site clearance is often necessary, adding to the development cost. The surrounding area may be of poor environmental quality.

Burgess model another name for ◊concentric-ring theory.

business park low-density office development of a type often established by private companies on ◊greenfield sites. The sites are often landscaped to create a pleasant working environment. Business parks tend to be located near motorway junctions and may have a high proportion of high-tech firms.

Business parks were introduced into the UK in the early 1980s, and by 1991 there were about 800 throughout the country.

bustee alternative spelling of ***basti***, an Indian name for a ◊shanty town.

butte steep-sided flat-topped hill. Buttes are remnants of eroded plateaus and are usually formed in horizontally layered sedimentary rocks, largely in arid areas (areas of low rainfall) such as Colorado, Utah, and Arizona in the USA. A large butte with a pronounced table-like profile is a ***mesa***.

C

caldera a very large basin-shaped ◊crater. Calderas are found at the tops of volcanoes, where the original peak has collapsed into an empty chamber beneath. The basin, many times larger than the original volcanic vent, may be flooded, producing a crater lake, or the flat floor may contain a number of small volcanic cones, produced by volcanic activity after the collapse.

Typical calderas are Kilauea, Hawaii; Crater Lake, Oregon, USA; and the summit of Olympus Mons, on Mars. Some calderas are wrongly referred to as craters, such as Ngorongoro, Tanzania.

canyon (Spanish *cañon* 'tube') deep, narrow valley or gorge running through mountains. Canyons are formed by stream down-cutting, usually in arid areas, where the stream or river receives water from outside the area.

There are many canyons in the western USA and in Mexico, for example the Grand Canyon of the Colorado River in Arizona, the canyon in Yellowstone National Park, and the Black Canyon in Colorado.

CAP abbreviation for ◊Common Agricultural Policy.

capital in a country, the city where the government headquarters are. The capital is usually the most important and largest city in a country; for example, London. Some countries have moved the seat of government to reduce strain on the largest city's infrastructure; for example, Brasilia is the specially built capital of Brazil.

carbonation form of ◊chemical weathering caused by rainwater that has absorbed carbon dioxide from the atmosphere and formed a weak carbonic acid. The slightly acidic rainwater is then capable of dissolving certain minerals in rocks. ◊Limestone is particularly vulnerable to this form of weathering.

cartography art and practice of drawing ◊maps.

cash crop crop grown solely for sale rather than for the farmer's own use, for example, coffee, cotton, or sugar beet. Many developing countries grow cash crops to meet their debt repayments rather than grow food for their own people. The price for these crops depends on financial interests, such as those of the multinational companies and the International Monetary Fund.

In Britain, the most widespread cash crop is the potato.

catch crop crop such as turnip that is inserted between two principal crops in a rotation in order to provide some quick livestock feed or soil improvement at a time when the land would otherwise be lying idle.

catchment area another name for ◊drainage basin, the area from which water is collected by a river and its tributaries. In human geography the term may be used to denote the area from which people travel to obtain a particular good or service (see ◊sphere of influence), such as the area from which a school draws pupils.

cave roofed-over cavity in the Earth's crust usually produced by the action of underground water or by waves on a seacoast (◊coastal erosion). Caves of the former type commonly occur in areas underlain by limestone, such as Kentucky, USA, and many Balkan regions, where the rocks are soluble in water. Celebrated caves include the Mammoth Cave in Kentucky, 6.4 km long and 38 m high; the Cheddar caves, England; Fingal's Cave, Scotland, which has a range of basalt columns; and Peak Cavern, England.

cavitation erosion of rocks caused by the forcing of air into cracks. Cavitation results from the pounding of waves on the coast and the swirling of turbulent river currents, and exerts great pressures, eventually causing rocks to break apart.

The process is particularly common at waterfalls, where the turbulent falling water contains many air bubbles, which burst to send shock waves into the rocks of the river bed and banks.

CBD abbreviation for ◊central business district.

census official gathering of information about the population in a particular area. The data collected are used by government departments in planning for the future in such areas as health, education, transport, and housing.

Most countries have a census of some sort. In the UK, a census has been conducted every ten years since 1801. Although the information about individual households remains secret for 100 years, data are available on groups of households down to about 200 (an enumeration district), showing such characteristics as age and sex structure, employment, housing types, car ownership, and qualifications held. The larger-scale information on population numbers, movements, and origins is published as a series of reports by the Office of Population Censuses and Surveys. The most recent census took place on 21 April 1991.

central business district (CBD) area of a town or city where most of the commercial activity is found. This area is dominated by shops, offices, entertainment venues, and local-government buildings. Usually the CBD is characterized by high rents and rates, tall buildings, and chain stores, and is readily accessible to pedestrians. It may also occupy the historic centre of the city and is often located where transport links meet.

central place or ***service centre*** place to which people travel from the surrounding area (◊hinterland) to obtain various goods or services. Central places will often be towns or areas within towns; for example, a shopping arcade that serves people in the immediate neighbourhood with low-order goods (see ◊hierarchy). Central places vary in importance; higher-order goods and services are provided where the ◊threshold population is reached.

According to ***central-place theory***, if each settlement of a particular order acts as a central place for certain levels of goods or services, there should be a regular pattern and distribution of settlements within an area. This theory was first put forward by German geographer Walter Christaller in 1903. He suggested that each settlement would be surrounded by a hexagonal sphere of influence (hexagonal rather than circular because circles cannot fit together exactly). The size of these hexagons depends on the order of the central place – village, town, or city. Each order would have a market area three times that of the settle-

central place

central-place theory: each central place is surrounded by a hexagonal sphere of influence

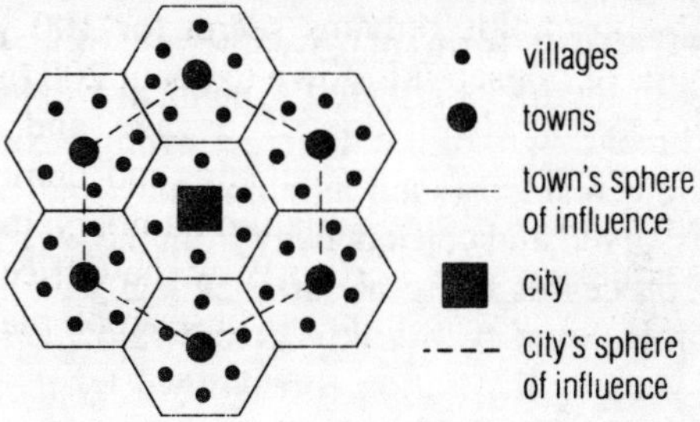

ment below. Settlements of each order would therefore be spaced at regular intervals in a spatial hierarchy.

Christaller took a number of factors for granted; for example, he assumed that transport was equally possible in all directions. In the real world this is not the case; however, the theory provides a starting point for explaining settlement distribution.

cereal grass grown for its edible, nutrient-rich, starchy seeds. The term refers primarily to wheat, oats, rye, barley, and triticale (a cross between wheat and rye), but may also refer to maize (corn), millet, sorghum, and rice. In 1984, world production exceeded 2 billion tonnes – enough to feed the world's population if supplies were evenly distributed.

chalk soft, fine-grained, whitish rock composed of calcium carbonate $CaCO_3$, extensively quarried for use in cement, lime, and mortar, and in the manufacture of cosmetics and toothpaste. ***Blackboard chalk*** in fact consists of gypsum (calcium sulphate, $CaSO_4$).

Chalk covers a wide area in Europe. In England it stretches in a belt from Wiltshire and Dorset continuously across Buckinghamshire and Cambridgeshire to Lincolnshire and Yorkshire, and also forms the North and South Downs, and the cliffs of S and SE England.

channel efficiency measure of the ability of a river channel to discharge water. Channel efficiency can be assessed by calculating the channel's ◊hydraulic radius. The most efficient channels are generally semicircular in cross-section, and it is this shape that water engineers try to create when altering a river channel to reduce the risk of ◊flooding.

chemical weathering form of ◊weathering brought about by a chemical change in the rocks affected. Chemical weathering involves the 'rotting', or breakdown, of the minerals within a rock, and usually produces a claylike residue (such as china clay and bauxite). Some chemicals are dissolved and carried away from the weathering source.

A number of processes bring about chemical weathering, such as ◊carbonation (breakdown by weakly acidic rainwater), ◊hydrolysis (breakdown by water), ◊hydration (breakdown by the absorption of water), and ◊oxidation (breakdown by the oxygen in water).

china clay or ***kaolin*** white clay formed by the chemical weathering of feldspar, a mineral found in ◊granite. China clay is used in making porcelain and as a filler in paper making and paints. economically important in the ceramic and paper industries. It is mined in the USA, France, Germany, and the UK, near St Austell, Cornwall.

chinook warm dry wind that blows downhill on the eastern side of the Rocky Mountains, North America. It often occurs in winter and spring when it produces a rapid thaw, and so is important to the agriculture of the area.

The chinook is similar to the ◊föhn in the valleys of the European Alps.

choropleth map map on which the average numerical value of some aspect of an area (for example, unemployment by county) is indicated by a scale of colours or ◊isoline shadings. An increase in average value is normally shown by a darker or more intense colour or shading. Choropleth maps are visually impressive but may mislead by suggesting sudden changes between areas.

cirque French name for a ◊corrie, a steep-sided hollow in a mountainside.

city important, or high-order, urban settlement. In the past, a town in Britain needed either a cathedral or a royal charter before it could be

choropleth

choropleth map showing the natural increase in population in North and South America

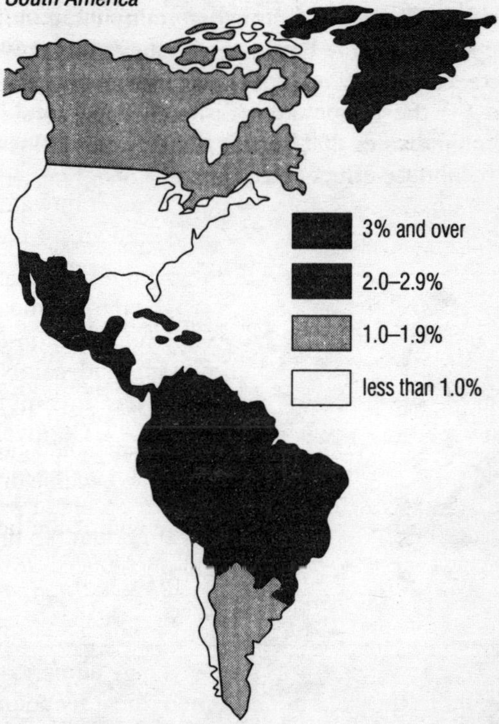

called a city, but in modern-day usage a city is a settlement with a population of more than 150,000. Royal charters are still awarded, however.

clay very fine-grained sedimentary deposit. When moistened it is becomes unstable, making it prone to ◊mass movement. On drying, it hardens to form an impermeable layer. It may be white, grey, red, yellow, blue, or black, depending on its composition.

Types of clay include alluvial clay and china clay. Clays have a variety of uses, some of which, such as pottery and bricks, date back to prehistoric times.

climate the average weather conditions of a particular place over a long period of time, usually 30 years. Several different climate zones may be identified. The main factors determining the variations of climate over the surface of the Earth are: (1) the effect of latitude and the tilt of the Earth's axis; (2) the large-scale movement of wind and ocean currents; and (3) the temperature difference between land and sea. Recent research indicates that human activity may influence world climate (see ◊greenhouse effect and ◊climatic change).

climate

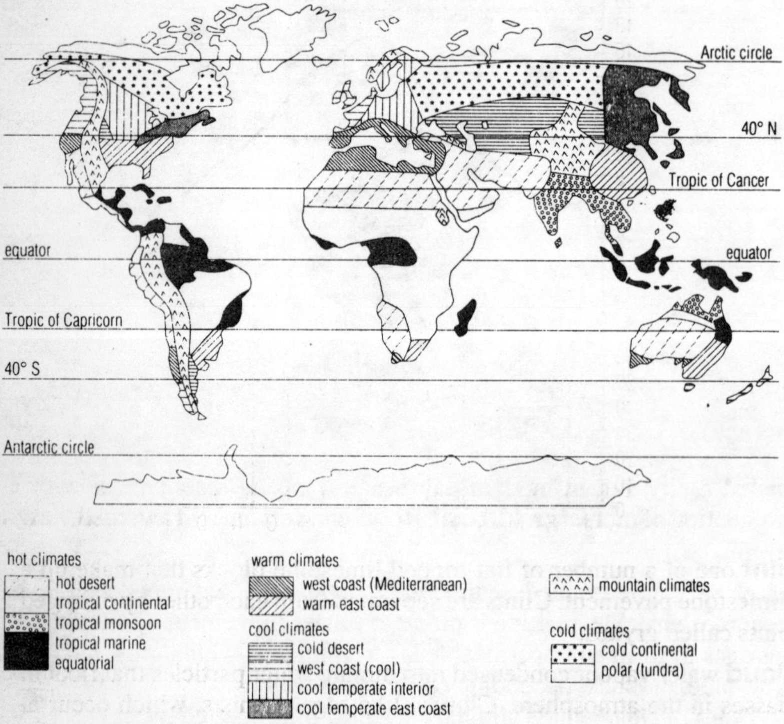

climatic change change in the climate of an area or of the whole world over an appreciable period of time.

Climatic changes have taken place regularly, most notably during the ◊Ice Age, when the climate of Scotland changed from arctic to temperate conditions 13,000 years ago. Modern climatic changes are being brought on by pollution changing the composition of the atmosphere and producing a ◊greenhouse effect.

climograph diagram that shows both the average monthly temperature and ◊precipitation of a place.

climograph

climograph of London

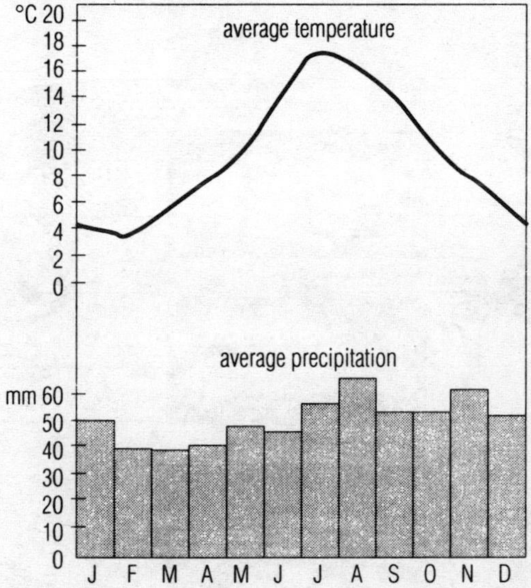

clint one of a number of flat-topped limestone blocks that make up a ◊limestone pavement. Clints are separated from each other by enlarged joints called grykes.

cloud water vapour condensed into minute water particles that float in masses in the atmosphere. Clouds, like fogs or mists, which occur at

cloud

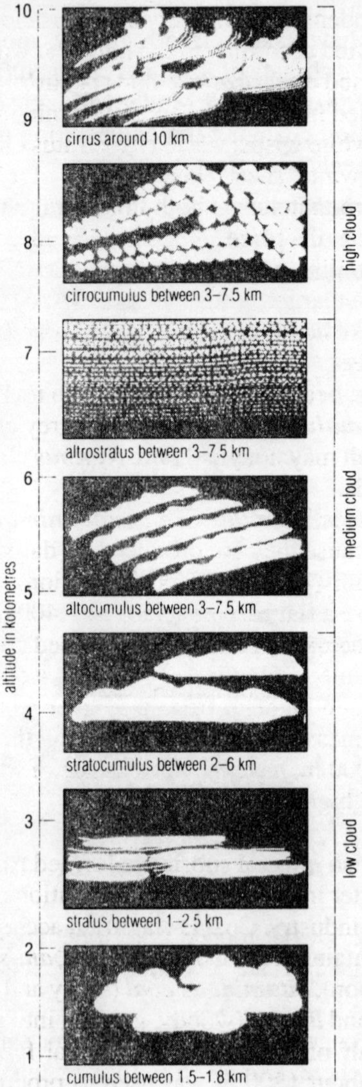
10
9
8
7
6
5
4
3
2
1
altitude in kolometres
cirrus around 10 km
cirrocumulus between 3–7.5 km
altrostratus between 3–7.5 km
altocumulus between 3–7.5 km
stratocumulus between 2–6 km
stratus between 1–2.5 km
cumulus between 1.5–1.8 km
high cloud
medium cloud
low cloud

lower levels, are formed by the cooling of air containing water vapour, which generally condenses around tiny dust particles.

Clouds are classified according to the height at which they occur and their shape. ***Cirrus*** and ***cirrostratus*** clouds occur at 10,000 m. The former, sometimes called mares'-tails, consist of minute specks of ice and appear as feathery white wisps, while cirrostratus clouds stretch across the sky as a thin white sheet. Three types of cloud are found at 3,000–7,500 m: cirrocumulus, altocumulus, and altostratus. ***Cirrocumulus*** clouds occur in small or large rounded tufts, sometimes arranged in the familiar pattern called mackerel sky. ***Altocumulus*** clouds are similar, but larger, white clouds, also arranged in lines. ***Altostratus*** clouds are like heavy cirrostratus clouds and may stretch across the sky as a grey sheet.

The lower clouds, occurring at heights of up to 1,800 m, may be of two types. ***Stratocumulus*** clouds are the dull grey clouds that give rise to a leaden sky which may not yield rain. ***Nimbus*** clouds are dark-grey, shapeless rain clouds.

Two types of clouds, ***cumulus*** and ***cumulonimbus***, are placed in a special category because they are produced by daily ascending air currents, which take moisture into the cooler regions of the atmosphere. Cumulus clouds have a flat base generally at 1,400 m where condensation begins, while the upper part is dome-shaped and extends to about 1,800 m. Cumulonimbus clouds have their base at much the same level, but extend much higher, often up to over 6,000 m. Short heavy showers and sometimes thunder may accompany them. ***Stratus*** clouds, occurring below 1,000 m, have the appearance of sheets parallel to the horizon and are like high fogs.

coal black or blackish mineral substance formed from the compaction of ancient plant matter in tropical swamp conditions. It is used as a fuel and in the chemical industry. Coal is classified according to the proportion of carbon it contains. The main types are ◊***anthracite*** (shiny, with more than 90% carbon), ***bituminous coal*** (shiny and dull patches, more than 80% carbon), and ***lignite*** (woody, grading into peat, 70% carbon).

In the second half of the 18th century, coal fuelled the Industrial Revolution. From about 1800, it was used to produce coalgas for gas

coal mining

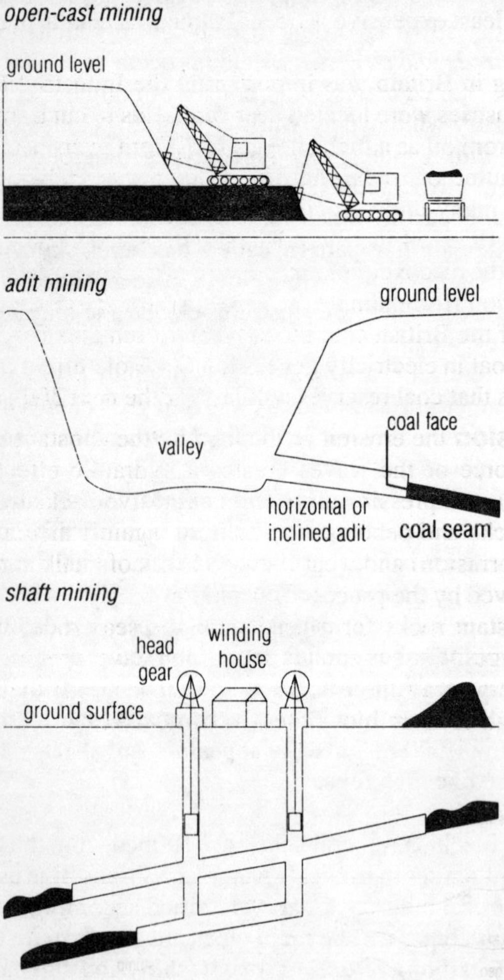

lighting, and coke for smelting iron ore. More recently it has been used by the petrochemical industry in the production of plastics. Coal burning is one of the main causes of ◊acid rain.

coal mining extraction of coal (a ◊sedimentary rock) from the Earth's crust. Coal mines may be opencast (see ◊opencast mining), ◊adit, or deepcast. The least expensive is opencast but this results in scars on the landscape.

Coal mining in Britain was important in the Industrial Revolution, and many industries were located near coalfields to cut transport costs. Competition from oil as a fuel, cheaper coal from overseas (USA, Australia), the decline of traditional users (town gas, railways), and the exhaustion of many underground workings resulted in the closure of mines (850 in 1955, 54 in 1992). Rises in the price of oil, greater productivity, and the discovery of new, deep coal seams suitable for mechanized extraction (for example, at Selby in Yorkshire) have improved the position of the British coal industry, but it remains very dependent of the use of coal in electricity generation (74% of current use). British Coal estimates that coal reserves will last for the next 300 years.

coastal erosion the erosion of the land by the constant battering of waves. The force of the waves creates a hydraulic effect (see ◊hydraulic action), compressing air to form explosive pockets in the rocks and cliffs. Rocks and pebbles may be flung against the cliff face (the process of ◊corrasion) and wear it away. Cliffs of chalk and limestone may be dissolved by the process of ◊solution.

Where resistant rocks form headlands, the sea erodes the coast in successive stages. First it creates cracks in ◊cave openings and then gradually wears away the interior of the caves until their roofs are pierced through to form blowholes. In time, caves at either side of a

coastal erosion

eroded headland at low tide

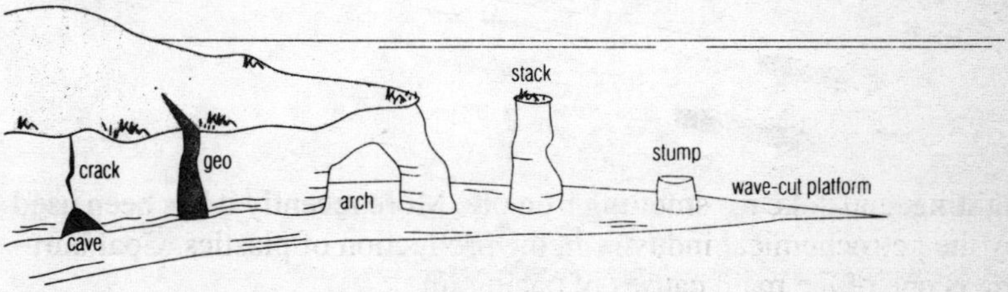

headland may unite to form a natural arch. When the roof of the arch collapses, a ◊stack is formed. This may be worn down further to produce a ◊stump and a ◊wave-cut platform.

coastal protection measures taken to prevent ◊coastal erosion. Many stretches of coastline are so severely affected by erosion that beaches are swept away, threatening the livelihood of seaside resorts, and buildings become unsafe.

To reduce erosion, several different forms of coastal protection may be employed. Structures such as sea walls attempt to prevent waves reaching the cliffs by deflecting them back to sea. Such structures are expensive and of limited success. A currently preferred option is to add sediment (beach nourishment) to make a beach wider. This causes waves to break early so that they have less power when they reach the cliffs. ◊Groynes may also be constructed to trap sediment and widen beaches.

coastal protection

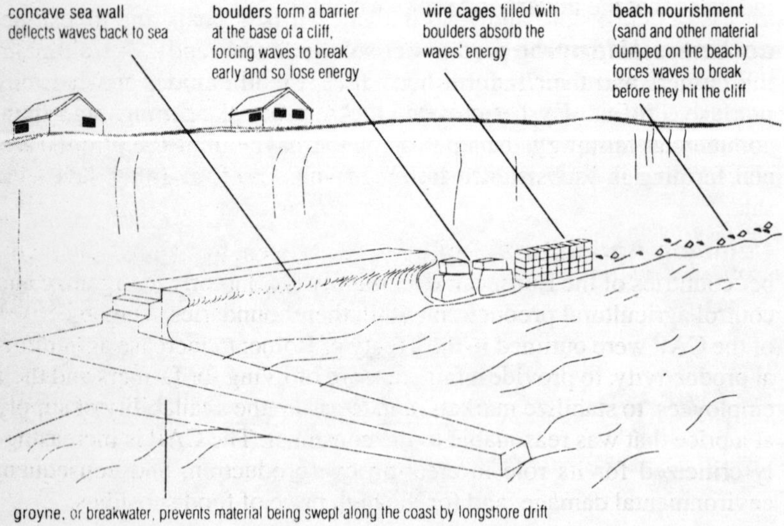

coffee tropical evergreen shrub whose beanlike seeds are roasted and ground to produce a drink of the same name. Coffee grows best on frost-free hillsides with moderate rainfall. The world's largest producers are Brazil, Colombia, and the Ivory Coast.

coke clean, light fuel produced when coal is strongly heated in an airtight oven. Coke contains 90% carbon and makes an useful domestic and industrial fuel (used, for example in the iron and steel industries and in the production of town gas).

collective farm farm in which a group of farmers pool their land, domestic animals, and agricultural implements, retaining as private property enough only for the members' own requirements. The profits of the farm are divided among its members. In ◊cooperative farming, farmers retain private ownership of the land.

The system was first developed in the USSR in 1917. China adopted collective farming from 1953, and Israel has a large number of collective farms, including the kibbutzes.

combe or ***coombe*** steep-sided valley found on the scarp slope of a chalk ◊escarpment. The inclusion of 'combe' in a placename usually indicates that the underlying rock is chalk.

commercial farming production of crops for sale and profit, although the farmers and their families may use a small amount of what they produce. Profits may be reinvested to improve the farm. Large-scale commercial farming is called ◊agribusiness; the opposite of commercial farming is ◊subsistence farming, where no food is produced for sale.

Common Agricultural Policy (CAP) system that allows the member countries of the European Community (EC) jointly to organize and control agricultural production within their boundaries. The objectives of the CAP were outlined in the Treaty of Rome: to increase agricultural productivity, to provide a fair standard of living for farmers and their employees, to stabilize markets, and to assure the availability of supply at a price that was reasonable to the consumer. The CAP is increasingly criticized for its role in creating overproduction, and consequent environmental damage, and for the high price of food subsidies.

At the heart of the CAP is a price support system, which guarantees prices for basic commodities; sets ◊quotas to prevent the overproduction of certain commodities, such as milk; and grants subsidies to encourage farmers to produce more of other commodities, such as oilseed rape.

common land unenclosed wasteland, forest, and pasture used in common by the community at large. Poor people throughout history have gathered fruit, nuts, wood, reeds, roots, game, and so on from common land; in dry regions of India, for example, the landless derive 20% of their annual income in this way, together with much of their food and fuel.

In the UK commons originated in the Middle Ages, when every manor had a large area of unenclosed, uncultivated land from which freeholders had rights to take the natural produce.

commune group of people or families living together, sharing resources and responsibilities.

commuter person who travels into a large town or city for work. For example, each working day more than 1.2 million people commute into London. A ***commuter belt*** is the area around a town in which commuters live (see also ◊dormitory town). Commuter settlements are generally more affluent than others, reflecting the wages of their residents, who prefer to live in a more attractive rural environment and can afford the daily costs of travelling to work.

Improved train services and better roads, especially motorways, have increased the number of commuters by making the countryside and towns more accessible.

comparison goods expensive goods, such as hi-fi systems and furniture, that the shopper will buy only after making a comparison between various models. A high ◊threshold population is needed to sustain a shop selling comparison goods, and people are prepared to travel some distance (◊range) to obtain them. Shops selling comparison goods often cluster in the central business district to share customers and increase trade.

compass any instrument for finding direction. The most commonly used is a magnetic compass, consisting of a thin piece of magnetic material with the north-seeking pole indicated, free to rotate on a pivot and mounted on a compass card on which the points of the compass are marked. When the compass is properly adjusted and used, the north-seeking pole will point to the magnetic north, from which true north can be found from tables of magnetic corrections.

Compasses not dependent on the magnet are gyrocompasses, dependent on the gyroscope, and radiocompasses, dependent on the use of radio. These are unaffected by the presence of iron, and are widely used in ships and aircraft.

component any separate part used by an ◊assembly industry to make the final product. Components can be the output from an industry as well as providing an input to another; for example, a motor car is made from several thousand individual components (windscreens, brake pedals, and so on), which are assembled into the finished product.

composite volcano steep-sided conical volcano, made up of alternate layers of ash and lava, formed at a ◊destructive margin. (By contrast, a ◊shield volcano is formed at a constructive margin.) The molten rock (magma) associated with composite volcanoes is very thick and often clogs up the vent. This can cause a tremendous buildup of pressure, which, once released, causes a very violent eruption. Examples of composite volcanoes are Mount St Helens in the USA and Mount Mayon in the Philippines.

concentric-ring theory or ***Burgess model*** hypothetical pattern of land use within an urban area, where different activities occur at different distances from the urban centre. The result is a sequence of rings. The theory was first suggested by the US sociologist E W Burgess in 1925. He said that towns expand outwards evenly from an original core so that each zone grows by gradual colonization into the next outer ring.

In addition, the cost of land may decrease with increased distance from the city centre as demand for it falls (see ◊bid-rent theory). This means that commercial activity that can afford high land values will be concentrated in the city centre.

concentric-ring theory

concentric model showing land use within and urban area

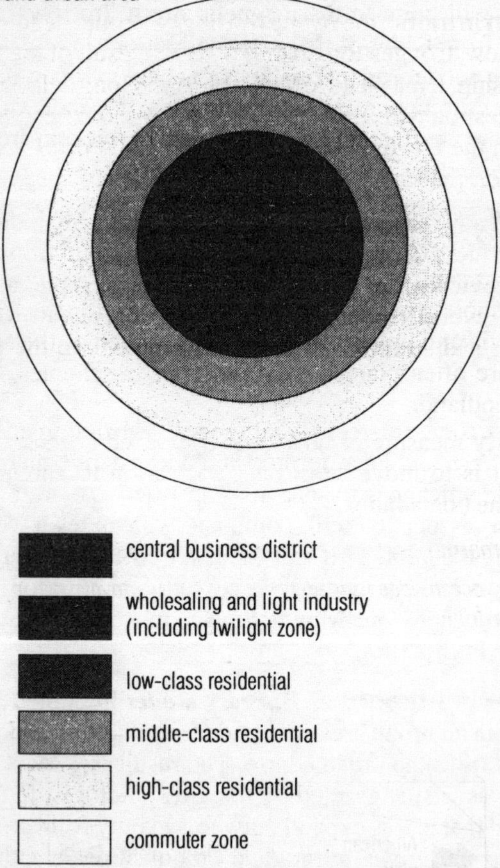

condensation conversion of a vapour to a liquid. It is the process by which water vapour turns into fine water droplets to form ◊cloud.

Condensation occurs when the air becomes completely saturated and is unable to hold any more water vapour. As air rises it cools and contracts – the cooler it becomes, the less water it can hold. The temperature at which the air becomes saturated is known as the ◊dew point. Condensation is an important part of the ◊hydrological cycle (water cycle).

confluence point at which two rivers join.

congestion in traffic, the overcrowding of a route, leading to slow and inefficient flow. Congestion on the roads is a result of the large increase in car ownership. It may lead to traffic jams and long delays as well as pollution. Congestion within urban areas may also restrict ◊accessibility.

In 1991 in the UK there were about 10 million cars and lorries, and the congestion that they cause is estimated to cost between £2 billion and £5 billion a year.

coniferous forest forest consisting of evergreen trees such as pines and firs. Most conifers grow quickly and can tolerate poor soil, steep slopes, and short growing seasons. Coniferous forests are widespread in Scandinavia and upland areas of the UK such as the Scottish Highlands, and are often planted in ◊afforestation schemes. Conifers also grow in ◊woodland.

connectivity measure of how well connected a transport network is (how easy it is to move from one destination to another). A simple measure is the ◊beta index.

constructive margin

ocean ridge formed where two plates move away from each other

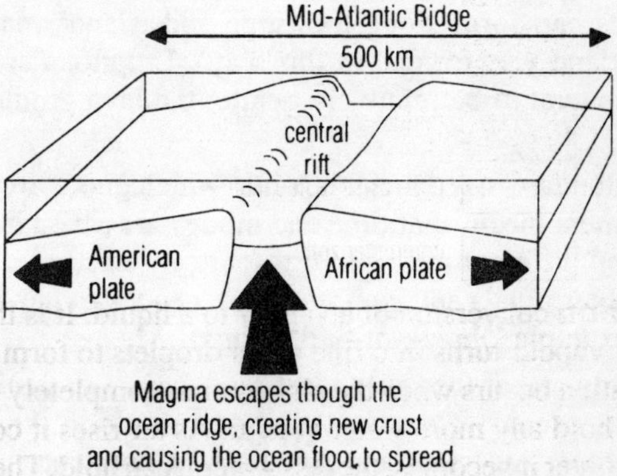

conservation action taken to protect and preserve the natural world, usually from pollution, overexploitation, and other harmful features of human activity. The late 1980s saw a great increase in public concern for the environment, with membership of conservation groups, such as Friends of the Earth, rising sharply. Globally the most important issues include the depletion of atmospheric ozone by the action of chlorofluorocarbons (CFCs), the build-up of carbon dioxide in the atmosphere (thought to contribute to an intensification of the ◊greenhouse effect), and the destruction of the tropical rainforests (see ◊deforestation).

In the UK the conservation debate has centred on water quality, road-building schemes, the safety of nuclear power, and animal rights.

conservative margin in plate tectonics, a region in which one plate slides past another. An example is the San Andreas Fault, California, where the movement of the plates is irregular and sometimes takes the form of sudden jerks, which cause the ◊earthquakes common in the San Francisco–Los Angeles area.

constructive margin in plate tectonics, a region in which two plates are moving away from each other. Magma, or molten rock, escapes to the surface along this margin to form new crust, usually in the form of a ridge. Over time, as more and more magma reaches the surface, the sea floor spreads – for example, the upwelling of magma at the Mid-Atlantic Ridge causes the floor of the Atlantic Ocean to grow at a rate of about 5 cm a year.

◊Volcanoes can form along the ridge and islands may result (for example, Iceland was formed in this way). Eruptions at constructive plate margins tend to be relatively gentle; the lava produced cools to form ◊basalt.

container standard-sized metal box into which goods are packed. The use of containers means that time and money are saved in transferring cargo between different means of transport; for example, from ship to train. Special terminals and equipment are needed to handle such containers (for example, ◊break-of-bulk points).

continent any one of the seven large land masses of the Earth, as distinct from the oceans. They are Asia, Africa, North America, South America, Europe, Australia, and Antarctica. Continents are constantly moving and evolving (see ◊plate tectonics). A continent does not end at the coastline; its boundary is the edge of the shallow continental shelf, which may extend several hundred kilometres out to sea.

continental drift theory that, about 200 million years ago, the Earth consisted of a single large continent (◊Pangaea) that subsequently broke apart to form the continents known today. The theory of continental drift was proposed in 1915 by the German meteorologist Alfred Wegener, but the means by which vast continental movements could take place were not satisfactorily explained until the study of ◊plate tectonics in the 1960s.

continental shelf gently sloping submarine plain extending into the ocean from a continent. The plain has a gradient of less than 1°; when the angle of the sea bed is 1°–5°, it is known as the ***continental slope***. This change usually occurs several hundred kilometres away from the land. The continental shelf around the UK contains large reserves of oil and gas.

contour line drawn on a map to join points of equal height. Contours are drawn at regular height intervals; for example, every 10 m. The closer together the lines are, the steeper the slope. Contour patterns can be used to interpret the relief of an area and to identify land forms.

conurbation or ***metropolitan area*** large continuous built-up area formed by the joining together of several urban settlements. Conurbations are often formed as a result of ◊urban sprawl. Typically, they have populations in excess of 1 million and some are many times that size; for example, the Osaka–Kobe conurbation in Japan, which contains over 16 million people.

convectional rainfall rainfall associated with hot climates, resulting from the uprising of convection currents of warm air. Air that has been warmed by the extreme heat of the ground surface rises to great heights and is abruptly cooled. The water vapour carried by the air condenses and rain falls heavily. Convectional rainfall is usually accompanied by a ◊thunderstorm.

Convectional rainfall occurs occasionally in the UK during hot summers.

convection current current caused by the expansion of a liquid or gas as its temperature rises. Convection currents arise in the atmosphere above warm land masses or seas, giving rise to ◊land breezes and ◊sea breezes respectively. Convection currents in the semisolid rock of the Earth's mantle are responsible for the movement of the rigid plates making up the Earth's surface (see ◊plate tectonics).

convenience good low-order product that is purchased frequently, such as stamps or bread. A ◊comparison good is a seldom purchased high-order product.

cooperative farming system in which individual farmers pool their resources (excluding land) to buy commodities such as seeds and fertilizers, and services such as marketing. It is a system of farming found throughout the world and is common in Denmark and countries of the former USSR. In ◊collective farming, land is also held in common.

core in physical geography, the innermost part of the Earth. It is divided into a solid inner core, the upper boundary of which is 1,700 km from the centre, and a semisolid outer core, 1,820 km thick. Both parts are thought to consist of iron and nickel. The temperature may be as high as 3,000°C.

Evidence for the nature of the core comes from seismology (observation of the paths and speeds of earthquake waves through the Earth), and calculations of the Earth's density.

core and periphery areas with different degrees of economic development. Within any particular region or country, development is unlikely to take place evenly. Areas with geographical advantages (such as soil fertility, raw materials, and access to trade routes) will become more developed than others. These are the ***core*** areas, where capital, ◊infrastructure, and employment are concentrated, leaving ***periphery*** areas that lack these resources. Core and periphery regions may be identified at many levels. On a national scale, for example, the UK has a northern periphery and southeast core.

Coriolis effect the effect of the the Earth's rotation on the atmosphere and on all objects on the Earth's surface. In the northern hemisphere it causes moving objects and currents to be deflected to the right; in the southern hemisphere it causes deflection to the left. The Coriolis effect

core and periphery

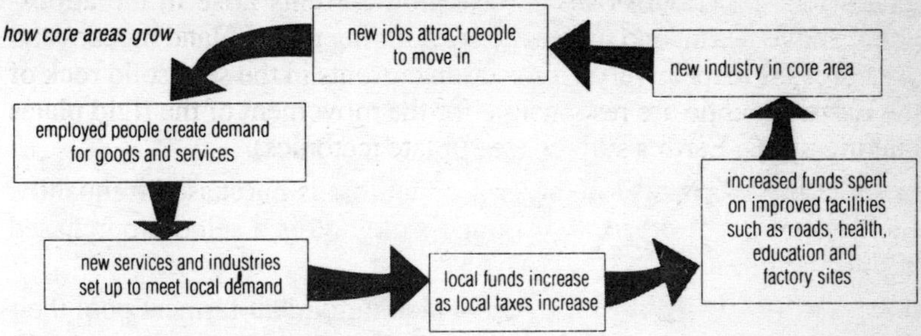

can be easily observed by watching water go down a plughole – it does not flow directly downwards but spins to the right (clockwise) or to the left (anticlockwise), depending on whether the observer is in the northern or southern hemisphere.

corn the main ◊cereal crop of a region – for example, wheat in the UK, oats in Scotland and Ireland, maize in the USA.

corrasion the grinding away of solid rock surfaces by particles carried by water, ice and wind. It is generally held to be the most significant form of ◊erosion. As the eroding particles are carried along they become eroded themselves due to the process of ◊attrition.

The term 'abrasion' is often used to mean the same thing, though, strictly speaking, abrasion refers to the *effect* of the process of corrasion.

correlation the relationship between two sets of information: they correlate when they vary together. If one set of data increases at the same time as the other, the relationship is said to be positive. If, as one set of data increases, the other decreases, the relationship is negative. Correlation can be shown by plotting a best-fit line on a ◊scatter diagram. See also ◊Spearman's rank correlation coefficient.

corrie (Welsh ***cwm***; French ***cirque***) steep-sided hollow in the mountainside of a glaciated area. A corrie is open at the front, and its sides and back are formed of ◊arêtes. There may be a lake in the bottom, called a tarn.

corrie

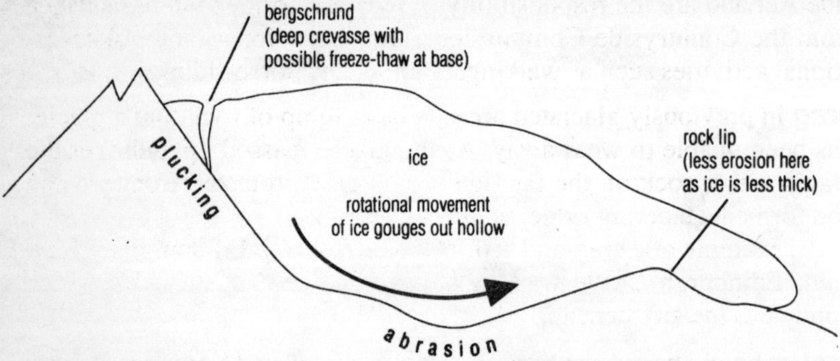

A corrie is formed as follows: (1) snow accumulates in a hillside hollow and turns to ice (enlarging the hollow by ◊nivation); (2) the hollow is deepened by ◊abrasion and ◊plucking; (3) the ice in the corrie rotates under the influence of gravity, deepening the hollow still further; (4) since the ice is thinner and moves more slowly at the foot of the hollow, a rock lip forms; (5) when the ice melts, a lake or tarn may be formed in the corrie. The steep back wall results from severe weathering by freeze-thaw, which provides material for further abrasion.

corrosion alternative name for ◊solution, the process by which water dissolves rocks such as limestone.

cotton tropical and subtropical plant. Fibres surround the seeds inside the ripened fruits, or bolls, and these are spun into yarn for cloth. Cotton production represents 5% of world agricultural output.

counterurbanization movement of people and employment away from urban areas to smaller towns and villages in rural locations. ◊Push factors within urban regions may be responsible – for example, ◊congestion, high land prices, and population pressure – together with ◊pull factors such as the perceived environmental quality of the countryside and improvements in transport systems.

country park pleasure ground or park, often located near an urban area, providing facilities for the public enjoyment of the countryside.

Country parks were introduced in the UK following the 1968 Countryside Act and are the responsibility of local authorities with assistance from the Countryside Commission. They cater for a range of recreational activities such as walking, boating, and horse-riding.

crag in previously glaciated areas, a large lump of rock that a glacier has been unable to wear away. As the glacier passed up and over the crag, weaker rock on the far side was largely protected from erosion and formed a tapering ridge, or ***tail***, of debris.

An example of a crag-and-tail feature is found in Edinburgh in Scotland; Edinburgh Castle was built on the crag (Castle Rock), which dominates the city beneath.

crater bowl-shaped depression, usually round and with steep sides. Craters are formed by explosive events such as the eruption of a volcano or by the impact of a meteorite. A ◊caldera is a much larger feature.

crevasse deep crack in the surface of a glacier; it can reach several metres in depth. Crevasses often occur where a glacier flows over the break of slope, because the upper layers of ice are unable to stretch and cracks result. Crevasses may also form at the edges of glaciers owing to friction with the bedrock.

croft small farm in the Highlands of Scotland, traditionally farming common land cooperatively. It is the only form of ◊subsistence farming found in the UK. Today, although grazing land is still shared, arable land is typically enclosed.

crop rotation system of regularly changing the crops grown on a piece of land. The crops are grown in a particular order to utilize and add to the nutrients in the soil and to prevent the build-up of insect and fungal pests. Including a legume crop, such as peas or beans, in the rotation helps build up nitrate in the soil because the roots contain bacteria capable of fixing nitrogen from the air.

A simple seven-year rotation, for example, might include a three-year ◊ley followed by two years of wheat and then two years of barley, before returning the land to temporary grass once more. In this way, the cereal crops can take advantage of the build-up of soil fertility that occurs during the period under grass.

crude birth rate the number of births per 1,000 of the population; see ◊birth rate.

crude death rate the number of deaths per 1,000 of the population; see ◊death rate.

crust the outermost part of the structure of the Earth, consisting of two distinct parts, the oceanic crust and the continental crust. The ***oceanic*** crust is on average about 10 km thick and consists mostly of basalt. By contrast, the ***continental*** crust is largely made of granite and is more complex in its structure. Because of the movements of ◊plate tectonics, the oceanic crust is in no place older than about 200 million years. However, parts of the continental crust are over 3 billion years old.

cuesta alternative name for ◊escarpment.

current the flow of a body of water or air, or of heat, moving in a definite direction; see ◊ocean current.

cwm Welsh name for a ◊corrie.

cyclone alternative name for a ◊depression, an area of low atmospheric pressure. A severe cyclone that forms in the tropics is called a tropical cyclone.

D

dairying the business of producing milk and milk products.

In the UK and the USA, over 70% of the milk produced is consumed in its liquid form, whereas areas such as the French Alps and New Zealand rely on easily transportable milk products such as butter, cheese, and condensed and dried milk. It is now usual for dairy farms to concentrate on the production of milk and for factories to take over the handling, processing, and distribution of milk as well as the manufacture of dairy products.

In Britain, the Milk Marketing Board (1933), to which all producers must sell their milk, forms a connecting link between farms and factories. Overproduction of milk in the EC has led to the introduction of quotas, which set a limit on the amount of milk for which a farmer may be paid.

dam structure built to hold back water in order to prevent flooding, provide water for irrigation and storage, and to provide hydroelectric power.

death rate the number of deaths per 1,000 of the population of an area over the period of a year. Death rate is a factor in ◊demographic transition.

Death rate is linked to a number of social and economic factors such as standard of living, diet, and access to clean water and medical services. The death rate is therefore lower in wealthier countries; for example, in the USA it is 9/1,000; in Nigeria 18/1,000.

debt something that is owed by a person, organization, or country, usually money, goods, or services. The ***national debt*** of a country is the total money owed by the national government to private individuals and banks. ***International debt*** is the money owed by one country to another.

The danger of the current scale of international debt (the so-called ***debt crisis***) is that the debtor country can only continue to repay its existing debts by means of further loans; for the Western countries, there is the possibility of a confidence crisis causing a collapse of the banking system.

decentralization the dispersion of a population away from a central point. A common form is ◊counterurbanization (in developed countries, the movement of industries and people away from cities). Examples in the UK include the move of the Department of Social Security to Newcastle and the Driver and Vehicle Licensing Authority (DVLA) to Swansea.

deciduous forest woodland area consisting of broad-leaved trees (such as oak) which shed their leaves in winter to reduce water loss and conserve energy. They are the natural vegetation of northern mainland Europe and the British Isles, but have been chopped down to make way for farming, industry, and settlement. Broad-leaved trees grow slowly, reaching maturity 100–200 years after being planted, thus limiting their economic value. Deciduous forest is contrasted with ◊coniferous forest.

deforestation the cutting down of forest without planting new trees to replace those lost (reafforestation) or allowing the forest to regenerate itself naturally. In tropical forests, such as those in the Amazon basin in South America, deforestation has been severe over the last few decades because of pressures from farmers and developers. Trees have been cut down to provide firewood and building materials and to make way for mining and urban developments.

Many people are concerned about the rate of deforestation as great damage is being done to the habitats of plants and animals. Deforestation also causes fertile soil to be blown away or washed into rivers, leading to ◊soil erosion and famine, and is thought to be partially responsible for the flooding of lowland areas (for example, in Bangladesh), because trees help to slow down water movement. It may also increase the carbon dioxide content of the atmosphere and intensify the ◊greenhouse effect, because there are fewer trees absorbing carbon dioxide from the air for photosynthesis.

delta

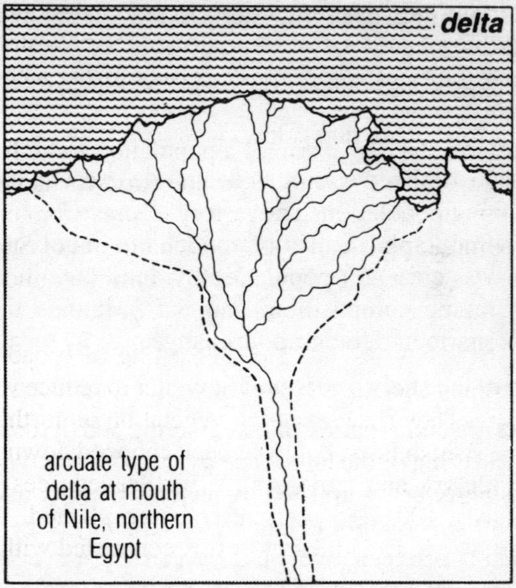

delta feature composed of silt formed when sediment is deposited at the mouth of a river, caused by the slowing of water on entering the sea. There are two main types of delta: an ***arcuate delta***, such as that of the Nile, is shaped like the Greek letter *delta* (giving rise to the name of the feature); a ***birdfoot delta***, like that of the Mississippi, is a seaward extension of the river's ◊levee system. Deltas are very fertile areas and are important for agriculture; for example, the Nile delta contains 90% of Egypt's farmland.

demographic transition any change in birth and death rates; over time, these generally shift from a situation where both are high to a situation where both are low. This may be caused by a variety of social factors (among them education and the changing role of women) and economic factors (such as higher standard of living and improved diet). The ***demographic transition model*** suggests that the change happens in four stages:

1) high birth rate, high death rate;

2) birth rate stays high, death rate starts to fall, giving maximum population growth;

3) birth rate starts to fall, death rate continues falling;

4) birth rate is low, death rate is low.

In some industrialized countries death rate exceeds birth rate, leading to a declining population. The history of many European countries follows the demographic transition model, but in poorer countries the pattern is far less clear. A ◊population pyramid (age–sex graph) illustrates demographic composition, and the ◊Malthus theory gives a worst-case scenario of demographic change.

demography the study of ◊population.

denudation the combination of ◊weathering and ◊erosion that results in a general lowering of the landscape over millions of years. The landscape is constantly being uplifted by tectonic activity and denuded by weathering and erosion. For example, the Himalayas are being formed by the collision between two plates, yet they are being constantly denuded by glacial erosion and frost shattering. As a result, overall uplift is not as great as it might be.

depressed area region with substandard economic performance, perhaps as a result of a change in industrial structure, such as a decline in manufacturing industry. An example in the UK is Clydeside, where traditional heavy industries have closed because of reduced demand and exhaustion of raw materials (such as coal and iron ore). Depressed areas may be characterized by high unemployment, low-quality housing, and poor educational standards. Government aid may be needed to reverse such decline (see ◊assisted area).

depression or *cyclone* region of low atmospheric pressure. Depressions form as warm air from the tropics spirals around cold polar air, producing cold and warm ◊fronts. The warm air, being less dense, rises above the cold air to produce the area of low pressure on the ground. Depressions bring unstable weather with cloud and rain (see ◊frontal rainfall).

depression

a typical depression showing low pressure at the centre

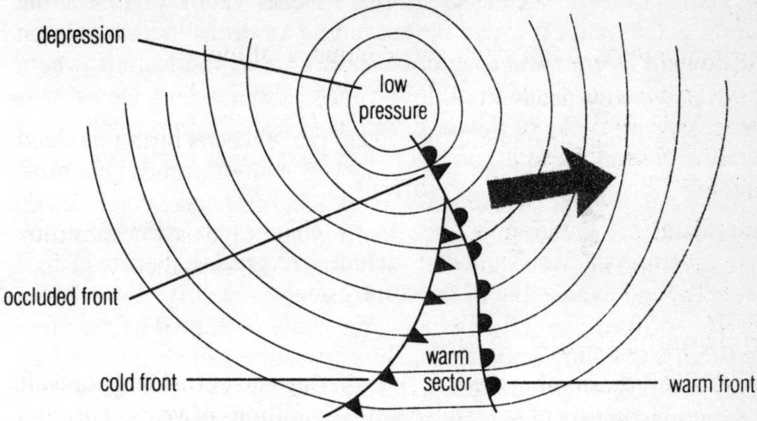

the fronts are associated with belts of rain (frontal rainfall)

In a deep depression the pressure at the centre is very much lower than that round about; it produces very strong winds, as opposed to a shallow depression in which the winds are lighter. Depressions tend to travel eastwards and can remain active for several days.

desalination removal of salt, usually from sea water, to produce fresh water for irrigation or drinking. Distillation has usually been the method adopted, but in the 1970s a cheaper process, using certain polymer materials that filter the molecules of salt from the water by reverse osmosis, was developed.

Desalination plants occur along the shores of the Middle East where fresh water is in short supply.

desert arid area without sufficient rainfall and, consequently, vegetation to support human life. The term includes the ice areas of the polar regions (known as cold deserts). Almost 33% of Earth's land surface is desert, and this proportion is increasing.

The ***tropical desert*** belts of latitudes from 5° to 30° are caused by the descent of air that is heated over the warm land and therefore has lost its moisture. Other natural desert types are the ***continental deserts***, such as the Gobi, that are too far from the sea to receive any moisture; ***rain-shadow deserts***, such as California's Death Valley, that lie in the lee of mountain ranges, where the ascending air drops its rain only on the windward slopes; and ***coastal deserts***, such as the Namib, where cold ocean currents cause local dry air masses to descend. Desert surfaces are usually rocky or gravelly, with only a small proportion being covered with sand. Deserts can be created by changes in climate, or by the human-aided process of desertification.

desertification creation of deserts by changes in climate, or by human-aided processes. The latter include overgrazing, destruction of forest belts, and exhaustion of the soil by intensive cultivation without restoration of fertility – all of which are usually prompted by the pressures of an expanding population. Desertification can be reversed by special planting (marram grass, trees) and by the use of water-absorbent plastic grains, which, added to the sand, enable crops to be grown. About 135 million people are directly affected by desertification, mainly in Africa, the Indian subcontinent, and South America.

desire line one of a set of straight lines drawn on a map to link the origins and destinations of people journeying to obtain a particular ◊good or service. The longer the desire line, the higher the order and the range of the good or service. Desire lines indicate the catchment area of a settlement.

destructive margin in plate tectonics, a region in which two plates are moving towards one another. One plate (because it is more dense) is forced to dive below the other in what is called the ***subduction zone***. The descending plate melts to form a body of magma in a region called the Benioff zone. Magma will rise to the surface through cracks and faults to form ◊composite volcanoes. If the two plates consist of more buoyant continental crust, subduction will not occur. Instead, the crust will crumple to form young fold mountains such as the Himalayas. Destructive margins are responsible for the formation of many major features, including young fold mountains, ocean trenches, and island arcs. Volcanoes and ◊earthquakes are common along destructive margins.

destructive margin

ocean trench and island arc formed by two plates moving towards one another

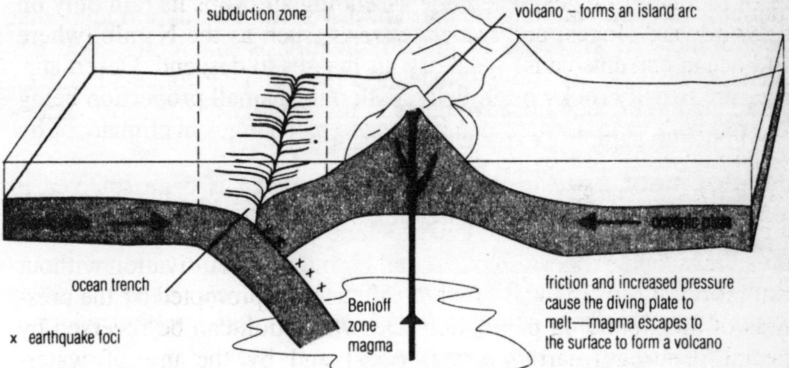

developed world or ***First World*** or ***the North*** the countries that have a money economy and a highly developed industrial sector. They generally also have a high degree of urbanization, a complex communications network, high ◊gross domestic product (over US $2,000) per person, low birth and death rates, high energy consumption, and a large proportion of the workforce employed in manufacturing or service industries (secondary to quaternary ◊industrial sectors). The developed world includes the USA, Canada, Europe, Japan, Australia, and New Zealand.

developing world or ***Third World*** or ***the South*** countries with a largely subsistence economy where the output per person and the average income are both low. These countries typically have low life expectancy, high birth and death rates, poor communications and health facilites, low literacy levels, high national debt, and low energy consumption per person. The developing world includes much of Africa and parts of Asia and South America. Terms like 'developing world' and 'less developed countries' are often criticized for implying that a highly industrialized economy (as in the ◊developed world) is a desirable goal.

development the acquisition by a society of industrial techniques and technology. It has led to the classification of the 'developed' nations of the First and Second Worlds and the poorer, 'developing' or 'underdeveloped' nations of the Third World. The assumption that development in the sense of industrialization is inherently good has been increasingly questioned since the 1960s, and the concept has been broadened to include improvements in the 'quality of life' – for example, in health care, life expectancy, education, and housing.

development area in the UK, a region that has been selected to receive government aid because of its poor economic performance; now called ◊assisted area.

dew precipitation in the form of moisture that collects on the ground. It forms after the temperature of the ground has fallen below the dew point of the air in contact with it. As the temperature falls during the night, the air and its water vapour become chilled, and condensation takes place on the cooled surfaces.

When moisture begins to form, the surrounding air is said to have reached its dew point. If the temperature falls below freezing point during the night, the dew will freeze, or if the temperature is low and the dew point is below freezing point, the water vapour condenses directly into ice; in both cases hoar frost is formed.

dew point temperature at which the air becomes saturated with water vapour. At temperatures below the dew point, the water vapour condenses out of the air as droplets. If the droplets are large they become deposited on the ground as dew; if small they remain in suspension in the air and form mist or fog. If the temperature falls below freezing point during the night, the dew will freeze to form hoar frost.

dip the angle and direction in which a ◊bed of rock is plunging. Rocks that are dipping have usually been affected by folding (see ◊fold).

discharge in a river, the volume of water passing a certain point per unit of time. It is usually expressed in cubic metres per second (cumecs). The discharge of a particular river channel may be calculated by multiplying the channel's cross-sectional area (in square metres) by the velocity of the water (in metres per second).

In the UK most small rivers have an average discharge of less than 1 cumec. The highest discharge usually occurs in early spring when melted snow combines with rainfall to increase the amount of water entering a river. Discharges are studied carefully by scientists as they can signal when a river is in danger of ◊flooding. Channel efficiency is an important factor in determining the likelihood of flooding at times of high discharge.

dispersed settlement settlement made up of buildings scattered over a wide area. Other settlement patterns include ◊nucleated settlements and ◊linear developments.

distance the space between two points. Distance is normally measured using centimetres, metres, and kilometres, but may also be looked at in terms of the time, cost, and inconvenience that travelling between places entails.

distance decay the effect that distance has on movement. As people's distance from a point increases, the amount of movement to that point generally decreases (a negative ◊correlation). Distance is a barrier to movement because of increased cost, inconvenience, and time, and limited knowledge about an area. The location of a firm's customers shows the effect of distance decay, whereas the destination of British tourists does not, since more visit Spain or Greece than the UK's own coastal resorts.

distributary river that has branched away from a main river. Distributaries are most commonly found on a ◊delta, where the very gentle gradient and large amounts of silt deposited encourage channels to split.

Channels are said to be ***braided*** if they branch away from a river and rejoin it at a later point.

diurnal temperature the range in temperature over a 24-hour period.

diversification in agriculture and business, the development of distinctly new products or markets. A company or farm may diversify in order to spread its risks or because its original area of operation is becoming less profitable. In the UK agricultural diversification has included offering accommodation and services to tourists – for example, bed and breakfast, camping and caravanning sites, and pony trekking.

distance

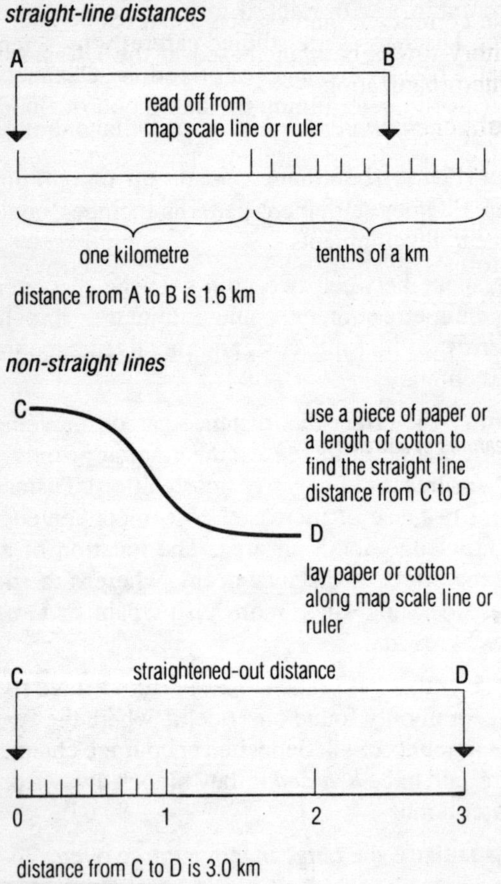

doldrums area of low atmospheric pressure along the equator, in the ◊intertropical convergence zone. The doldrums are characterized by calm or very light westerly winds, during which there may be sudden squalls and stormy weather. For this reason the areas are avoided as far as possible by sailing ships.

dormitory town rural settlement that has a high proportion of ◊commuters in its population. The original population may have been displaced by these commuters and the settlements enlarged by housing estates. Dormitory towns have increased in the UK since 1960 as a result of ◊counterurbanization.

drainage basin or ***catchment area*** the area of land drained by a river and its tributaries. The edge of a drainage basin is called the watershed.

drought period of prolonged dry weather. The area of the world subject to serious droughts, such as the Sahara, is increasing because of destruction of forests, overgrazing, and poor agricultural practices.

In the UK, drought is defined as the passing of 15 days with less than 0.2 mm of rain.

drainage basin

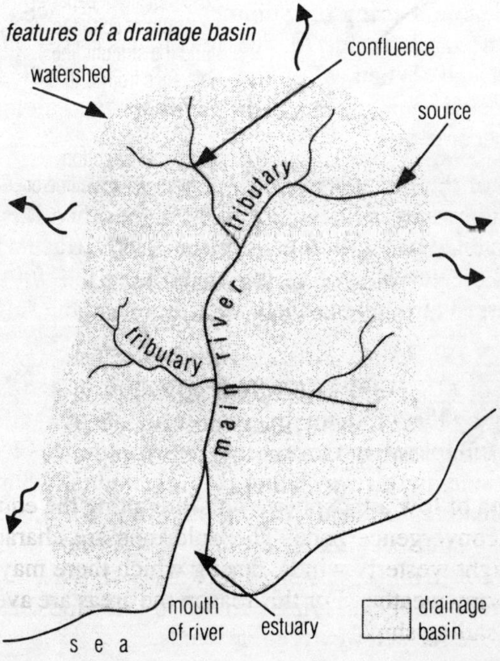

drumlin long streamlined hill created by ◊glacial deposition. Rocky debris (moraine) is gathered up by the glacial icesheet and moulded to form a low egg-shaped mound, 8–60 m in height and 0.5–1 km in length. Drumlins commonly occur in groups on the floors of glacial troughs, producing a 'basket-of-eggs' landscape. They are important indicators of the direction of ice flow, as their blunt ends point upstream, and their gentler slopes trail off downstream.

dry valley valley without a river at its bottom. Such valleys are common on the dip slopes of chalk ◊escarpments, and were probably formed by rivers. However, chalk is permeable (water passes through it) and so cannot retain surface water. How, therefore, could rivers have passed over its surface? Two popular theories have arisen to explain how this might have taken place:

1) During the last ice age the chalk might have frozen and been rendered impermeable. During the summer thaw, water would then have flowed over the land, unable to sink into it, and river valleys would have been formed. When, after the ice age, the chalk thawed and became permeable again, rivers could no longer flow along the valleys and so these became dry.

2) At the end of the last ice age so much meltwater might have been created that the ◊water table would be far higher than it is today. This would have enabled water to flow over the chalk surface without being absorbed, and create valleys. As the water table fell with time, however, water passed through the chalk once more and the valleys became dry.

dune mound or ridge of wind-drifted sand. Loose sand is blown and bounced along by the wind, up the windward side of a dune. The sand particles then fall to rest on the lee side, while more are blown up from the windward side. In this way a dune moves gradually downwind.

Dunes are features of sandy deserts and beach fronts. The typical crescent-shaped dune is called a ***barchan***. It is formed in sandy desert, where winds blow from a constant direction. ***Seif dunes*** are longitudinal and form on bare rocks, lying parallel to the wind direction. ***Star-shaped dunes*** are formed by irregular winds.

dust bowl area in the Great Plains region of North America (Texas to Kansas) that suffered extensive wind erosion as the result of drought and poor farming practice in once fertile soil. Much of the topsoil was blown away in the droughts of the 1930s and the 1980s.

Similar dust bowls are being formed in many areas today, noticeably across Africa, because of overcropping and overgrazing.

dyke sheet of ◊igneous rock created by the intrustion of magma (molten rock) across layers of pre-existing rock. It may form a ridge when exposed on the surface. It should not be confused with a ◊sill, which is intruded *between* layers of rock.

The term also refers to an embankment built along coastlines to prevent the flooding of lowland coastal

E

Earth third planet from the Sun. It is almost spherical, flattened slightly at the poles. 70% of the surface (including the north and south polar icecaps) is covered with water. The Earth is surrounded by a life-supporting atmosphere and is the only planet on which life is known to exist.
structure The Earth's interior is thought to be composed of a number of concentric layers: an inner ◊core of solid iron and nickel; an outer core of molten iron and nickel; and a ◊mantle of mostly solid rock, separated by the ◊Mohoroviçić discontinuity from the Earth's ◊crust. Evidence for the layered structure has been gathered by scientists surveying the paths taken by seismic waves (earthquake waves), which travel at different speeds through different materials. The crust and the topmost layer of the mantle (the lithosphere) form about 12 large moving plates, some of which carry the continents. The plates are in constant, slow motion, called tectonic drift (see ◊plate tectonics).

earthquake shaking of the Earth's surface as a result of the sudden release of stresses built up in the Earth's crust. The study of earthquakes is called ***seismology***.

Most earthquakes occur along faults (fractures or breaks) in the crust. ◊Plate tectonic movements generate the major proportion: as two plates move past each other they can become jammed and deformed, and a series of shock waves (seismic waves) occur when they spring free. Their force is measured on the ◊Richter scale, and the effect of an earthquake is measured on the Mercalli scale. The point at which an earthquake originates is the ***focus***; the point on the Earth's surface directly above this is the ***epicentre***.

Most earthquakes happen at sea and cause little damage. However, when severe earthquakes occur in highly populated areas they can cause great destruction and loss of life. A reliable form of earthquake prediction has yet to be developed, although the ◊seismic-gap theory has had some success in identifying likely locations.

The San Andreas fault in California, where the North American and Pacific plates move past each other, is the site of many earthquakes. It is part of the 'ring of fire', the belt of earthquakes and volcanoes circling the Pacific.

ecology (Greek *oikos* 'house') study of the relationship among organisms and the environments in which they live, including all living and nonliving components. The term was coined by the biologist Ernst Haeckel 1866.

Ecology may be concerned with individual organisms (for example, behavioural ecology, feeding strategies), with populations (for example, population dynamics), or with entire communities (for example, competition between species for access to resources in an ecosystem, or predator–prey relationships). A knowledge of ecology is important in addressing many environmental problems, such as the consequences of pollution.

economy of scale the reduction in costs per item (unit costs) that results from large-scale production. The high capital costs of machinery or a factory are spread across a greater number of units as more are produced. This may be a result of automation or ◊mass production; for example, in the car industry. Economies of scale can also be produced when firms that need similar services locate together, sharing the costs of their services; for example, on industrial estates.

ecosystem ecological unit made up of living organisms and the non-living, or physical, environment with which they interact. Ecosystems can be identified at different scales – for example, the global ecosystem consists of all the organisms living on Earth, the Earth itself (both land and sea), and the atmosphere above; a freshwater pond ecosystem consists of the plants and animals living in the pond, the pondwater and all the substances dissolved or suspended in that water, and the rocks, mud, and decaying matter that make up the pond bottom.

Ecosystems are fragile because the relationship between the components of an ecosystem is so finely balanced: the alteration or removal of any one component can cause the whole ecosystem to collapse. The ◊rainforest is an example of an ecosystem under threat.

edge link on a ◊topological map. Nodes, or junctions, are formed where edges meet and intersect.

electrification introduction of electricity into an area or a facility (such as railways). In the UK many intercity rail routes have been electrified in recent years to improve efficiency and cut operating costs.

El Niño warm ◊ocean current recurring every 5–8 years or so in the E Pacific off South America. It warms the nutrient-rich waters along the coast of Ecuador and Peru, killing cold-water fishes and plants, and is an important factor in global weather.

El Niño is believed to be caused by the failure of trade winds and, consequently, of the cool ocean currents normally driven by these winds. Warm surface waters then flow in from the east.

The phenomenon can disrupt the climate of the area disastrously, and has played a part in causing famine in Indonesia in 1983, bush fires in Australia because of drought, rainstorms in California and South America, the destruction of Peru's anchovy harvest and wildlife 1982–1983, and numerous typhoons in Japan in 1991.

employment the paid jobs that people do. Employment may reflect the different ◊industrial sectors in which people work or the way in which the jobs are created (see ◊formal employment and ◊informal employment). Jobs are a source of wealth and an essential part of economic success within an area, triggering a ◊multiplier effect.

englacial within the body of a ◊glacier. Rocky material may become englacial either by falling onto the glacier surface, where it becomes buried by subsequent snowfall, or into a crevasse (a deep crack in the ice). Englacial material will eventually melt out of the glacier, usually at its snout, or foot, to form depositional features such as till and kames.

enterprise zone special zone designated by government to encourage industrial and commercial activity, usually in economically depressed areas. Investment is attracted by means of tax reduction and other financial incentives.

In the UK, enterprise zones were introduced 1980 to encourage regional investment in areas such as the depressed inner cities. Industrial and commercial property are exempt from rates, development land tax, and from certain other restrictions such as planning permission. The Isle of Dogs in London's docklands was extensively developed as an enterprise zone during the 1980s.

environment the conditions affecting a particular organism, including physical surroundings, climate, and influences of other living organisms. In common usage, 'the environment' often means the total global environment, without reference to any particular organism.

epicentre the point on the Earth's surface immediately above the focus of an ◊earthquake. Most damage usually takes place at an earthquake's epicentre.

equator imaginary line, 40,092 km long, encircling the broadest part of the Earth, and representing 0° latitude. It divides the Earth into two halves, called the northern and the southern hemispheres.

erosion wearing away of the Earth's surface, caused by the breakdown and transportation of particles of rock or soil (by contrast, ◊weathering does not involve transportation). Agents of erosion include the sea, rivers, glaciers, and wind. There are several processes of erosion including ◊hydraulic action, ◊corrasion, ◊attrition, and ◊solution.

erratic displaced rock that has been transported by a ◊glacier from one area to another. For example, in East Anglia, England, erratics have been found that have been transported from as far away as Scotland and Scandinavia.

escarpment or ***cuesta*** large ridge created by the erosion of folded rocks. It has one steep side (scarp) and one gently sloping side (dip). Escarpments are common features of chalk landscapes, such as the Chiltern Hills and the North Downs in England. Certain features are associated with chalk escarpments, including dry valleys (formed on the dip slope), combes (steep-sided valleys on the scarp slope), and springs.

escarpment

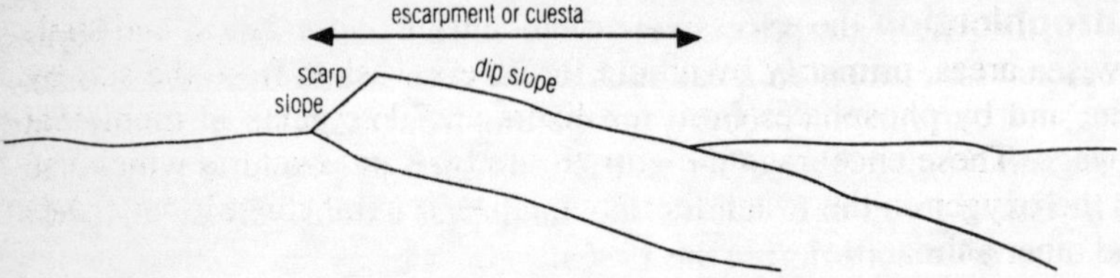

features associated with a chalk escarpment

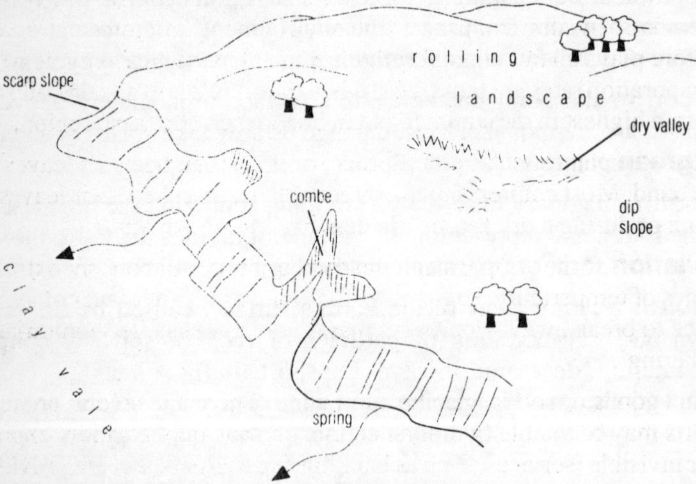

esker winding steep-walled ridge formed beneath a glacier. It is made of sands and gravels, and represents the course of a subglacial river channel. Eskers vary in height from 3 to 30 m and can be up to 160 km or so in length.

estuary river mouth widening into the sea, where fresh water mixes with salt water and tidal effects are felt.

eustatic change worldwide rise or fall in sea level caused by a change in the amount of water in the oceans (by contrast, ◊isostasy involves a rising or sinking of the land). During the last ice age sea level fell because water became 'locked up' in the form of ice and snow, and less water reached the oceans.

eutrophication the excessive enrichment of rivers, lakes, and shallow sea areas, primarily by nitrate fertilizers washed from the soil by rain, and by phosphates from fertilizers and detergents in municipal sewage. These encourage the growth of algae and bacteria which use up the oxygen in the water, thereby making it uninhabitable for fishes and other animal life.

evaporation process by which a liquid gradually turns to vapour. The evaporation of liquid water to form water vapour is responsible for the movement of water from the Earth's surface to the atmosphere, and therefore plays a key role in the ◊hydrological cycle (water cycle).

Evaporation rates are most rapid when the air is warm and dry, and are therefore highest in the summer and in hot ◊desert and ◊arid regions.

evergreen plant such as pine, spruce, or holly, that bears its leaves all year round. Most conifers are evergreen. Plants that shed their leaves in autumn or during a dry season are described as deciduous.

exfoliation form of ◊physical weathering brought about by extreme changes of temperature. Exfoliation may cause the outer layer of a rock surface to break away – a process that is sometimes called 'onion-skin weathering'.

export goods or service produced in one country and sold to another. Exports may be visible (goods, such as cars, that are physically exported) or invisible (services, such as banking and tourism, that are provided in the exporting country but paid for by residents of another country).

extensive agriculture ◊farming system where the area of the farm is large but there are low inputs (such as labour or fertilizers). Extensive farming generally gives rise to lower yields per hectare than ◊intensive agriculture. For example, in East Anglia, intensive use of land may give wheat yields as high as 53 tonnes per hectare, whereas an extensive wheat farm on the Canadian prairies may produce an average of 8.8 tonnes per hectare.

extrusive rock or ***volcanic rock*** ◊igneous rock formed on the surface of the Earth; for example, basalt. It is usually fine-grained, unlike the more coarse-grained ◊intrusive rocks (igneous rocks formed under the surface). The magma (molten rock) that cools to form extrusive rock may reach the surface through a crack, such as the constructive margin at the Mid Atlantic Ridge, or through the vent of a volcano.

Extrusive rock can be either ***lava*** (solidified magma, such as basalt) or a ***pyroclastic deposit*** (hot rocks or ash).

F

factory farming intensive rearing of poultry or animals for food, usually on high-protein foodstuffs in confined quarters. Chickens for eggs and meat, and calves for veal are commonly factory farmed. Some countries restrict the use of antibiotics and growth hormones as aids to factory farming, because they can persist in the flesh of the animals after they are slaughtered. Many people object to factory farming for moral as well as health reasons.

Egg-laying hens are housed in 'batteries' of cages arranged in long rows. In the UK in 1990, the number of factory-farmed table chickens numbered 600 million. Approximately 6 million free range chickens are reared each year.

fallow land ploughed and tilled, but left unsown for a season to allow it to recuperate; see ◊crop rotation. This is particularly common in developing-world farming systems (for example, ◊shifting cultivation) that do not have access to artificial fertilizers to maintain soil fertility.

family planning spacing or preventing the birth of children. Access to family-planning services is a significant factor in women's health as well as in limiting population growth (see ◊demographic transition). If all those women who wished to avoid further childbirth were able to do so, the number of births would be reduced by 27% in Africa, 33% in Asia, and 35% in Latin America; and the number of women who die during pregnancy or childbirth would be reduced by about 50%.

The average number of pregnancies per woman is two in the industrialized countries, compared to six or seven pregnancies per woman in the developing world. According to a World Bank estimate, doubling the annual $2 billion spent on family planning would avert the deaths of 5.6 million infants and 250,000 mothers each year.

famine severe shortage of food affecting a large number of people. Almost 750 million people (equivalent to double the population of Europe) worldwide suffer from hunger and malnutrition. Famines are usually explained as being caused by insufficient food supplies. Most Western famine-relief agencies, such as the international Red Cross, set out to supply food or to increase its local production, rather than becoming involved in local politics.

Famine may be the result of drought, as in Ethiopia in the 1980s, and of population pressure (as predicted by the ◊Malthus theory).

farming system type of agriculture; for example, ◊intensive agriculture and ◊extensive agriculture. The decision to practise a particular type of farming will depend upon the inputs available (nature of the land, labour, government subsidies, climatic and other factors), together with the the knowledge and abilities of the farmer.

faulting the cracking and displacement of adjacent blocks of rock as a result of the severe stresses set up by tectonic activity (see ◊plate tectonics).

Faults produce lines of weakness that are often exploited by processes of weathering and erosion. Coastal caves and geos (narrow inlets) often form along faults and, on a larger scale, rivers may follow the line of a fault.

The San Andreas Fault, California, marks a ◊conservative margin, where two tectonic plates slide past each other.

favela Brazilian makeshift housing or ◊shanty town.

fertilizer substance containing some or all of a range of about 20 chemical elements necessary for healthy plant growth, used to compensate for the deficiencies of poor or depleted soil. Fertilizers may be ***organic***, for example farmyard manure, composts, bonemeal, blood, and fishmeal; or ***inorganic***, in the form of compounds, mainly of nitrogen, phosphate, and potash, which have been used on a very much increased scale since 1945.

Because externally applied fertilizers tend to be in excess of plant requirements and drain away to affect lakes and rivers (see ◊eutrophication), attention has turned to the modification of crop plants themselves. Plants of the legume family, including the bean, clover, and lupin, live in symbiosis with bacteria located in root nodules, which fix nitrogen from the atmosphere. Research is now directed to producing a similar relationship between such bacteria and crops such as wheat.

fetch the distance of open water over which wind can blow to create ◊waves. The greater the fetch, the more potential power waves have when they hit the coast. In the south and west of England the fetch stretches for several thousand kilometres, all the way to South America. This combines with the southwesterly ◊prevailing winds to cause powerful waves and serious ◊coastal erosion along south- and west-facing coastlines.

fiord alternative spelling of ◊fjord.

firn or ***neve*** snow that has lain on the ground for a full calendar year. Firn is common at the tops of mountains; for example, the Alps in Europe. After many years, compaction turns firn into ice and a ◊glacier forms.

fish farming or ***aquaculture*** raising fish and shellfish (molluscs and crustaceans) under controlled conditions in tanks and ponds, sometimes in offshore pens. It has been practised for centuries in the Far East, where Japan alone produces some 100,000 tonnes of fish a year. In the 1980s one-tenth of the world's consumption of fish was farmed, notably carp, catfish, trout, salmon, turbot, eel, mussels, clams, oysters, and shrimp.

The 300 trout farms in Britain produce over 9,000 tonnes per year, and account for 90% of home consumption.

fishing the harvesting of fish and shellfish from the sea or from fresh water – for example, cod from the North Sea, and carp from the lakes of China and India. Fish are an excellent source of protein for humans, and fish products such as oils and bones are used in industry to produce livestock feed, fertilizers, glues, and drugs. The greatest proportion of the world's catch comes from the oceans.

Between 1950 and 1970, the world fish catch increased by an average of 7% each year. Refrigerated factory ships allowed filleting and processing to be done at sea, and Japan evolved new techniques for locating shoals (by sonar and radar) and catching them (for example, with electrical charges and chemical baits). By the 1970s, overfishing had led to serious depletion of stocks, and heated confrontations between countries using the same fishing grounds. A partial solution was the extension of fishing limits to 320 km around countries' shorelines. The North Sea countries have experimented with the artificial breeding of fish eggs and release of small fry into the sea. In 1988, overfishing of the NE Atlantic led to hundreds of thousands of starving seals on the northern coast of Norway. A United Nations resolution was passed 1989 to end drift-net fishing (an indiscriminate method).

fjord or ***fiord*** narrow sea inlet enclosed by high cliffs. Fjords are found in Norway, New Zealand, and western parts of Scotland. They are formed when an overdeepened ◊glacial trough is drowned by a rise in sea-level. At the mouth of the fjord there is a characteristic lip causing a shallowing of the water. This is due to reduced glacial erosion at this point.

flash flood flood of water in a normally arid area brought on by a sudden downpour of rain. Flash floods are rare and usually occur in mountainous areas. They may travel many kilometres from the site of the rainfall. Because of the suddenness of flash floods, little warning can be given of their occurrence. In 1972 a flash flood at Rapid City, South Dakota, USA, killed 238 people along Rapid Creek.

flint compact, hard, brittle mineral, commonly grey in colour, found in nodules in chalk deposits – for example, in the chalk downlands of southern England. Flint implements were widely used in prehistory.

Because of their hardness, flint splinters are used for abrasive purposes and, when ground into powder, added to clay during pottery manufacture. Flints have been used for making fire by striking the flint against steel, which produces a spark, and for discharging guns.

flooding the inundation of land that is not normally covered with water. Flooding from rivers commonly takes place after heavy rainfall or in the spring after winter snows have melted. The river's ◊discharge (volume of water carried in a given period) becomes too great, and water spills over the banks onto the surrounding flood plain. Small floods may happen once a year – these are called ***annual floods*** and are said to have a one-year return period. Much larger floods may occur on average only once every 50 years.

Flooding is least likely to occur in an efficient channel that is semi-circular in shape (see ◊channel efficiency). Flooding can also occur at the coast in stormy conditions (see ◊storm surge) or when there is an exceptionally high tide. The Thames Flood Barrier was constructed in 1982 to prevent the flooding of London from the sea.

flood plain area of periodic flooding that occurs inland along the course of river valleys. When river discharge exceeds the capacity of the channel, water rises over the channel banks and floods the adjacent low-lying lands. As water spills out of the channel some alluvium (silty material) will be deposited on the banks to form ◊levees (raised river banks). This water will slowly seep into the flood plain, depositing a new layer of rich fertile alluvium as it does so. Many important flood plains, such as the inner Niger delta in Mali, occur in arid areas where their exceptional productivity has great importance for the local economy. Flood plain features include ◊meanders and ◊oxbow lakes.

flood plain

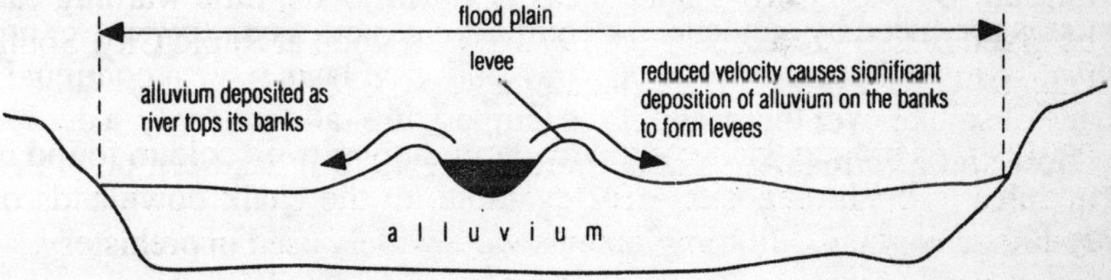

flue-gas desulphurization process of removing harmful sulphur pollution from gases emerging from a boiler. Sulphur compounds such as sulphur dioxide are commonly produced by burning ◊fossil fuels, especially coal in power stations, and are the main cause of ◊acid rain.

fluvioglacial of a process or landform, associated with glacial meltwater. Meltwater, flowing beneath or ahead of a glacier, is capable of transporting rocky material and creating a variety of landscape features, including eskers, kames, and outwash plains. The material transported will tend to be rounded and consist mainly of well-sorted sands and gravels.

focus the point within the Earth's crust at which an ◊earthquake originates. The point on the surface that is immediately above the focus is called the epicentre.

fog cloud that collects at the surface of the Earth, composed of water vapour that has condensed on particles of dust in the atmosphere. Cloud and fog are both caused by the air temperature falling below ◊dew point. The thickness of fog depends on the number of water particles it contains.

There are two types of fog. An ***advection fog*** is formed by the meeting of two currents of air, one cooler than the other, or by warm air flowing over a cold surface. Sea fogs commonly occur where warm and cold currents meet and the air above them mixes. A ***radiation fog*** forms on clear, calm nights when the land surface loses heat rapidly (by radiation); the air above is cooled to below its dew point and condensation takes place.

Officially, fog refers to a condition when visibility is reduced to 1 km or less, and mist or haze to that giving a visibility of 1–2 km. A mist is produced by condensed water particles, and a haze by smoke or dust. Industrial areas uncontrolled by pollution laws have a continual haze of smoke over them, and if the temperature falls suddenly, a dense yellow smog forms. At some airports since 1975 it has been possible

fog

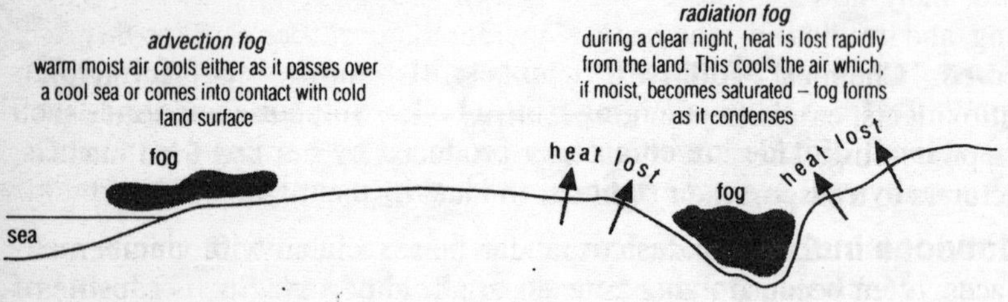

for certain aircraft to land and take off blind in fog, using radar navigation.

föhn or ***foehn*** warm dry wind that blows down the leeward slopes of mountains. The air heats up as it descends because of the increase in pressure, and it is dry because all the moisture was dropped on the windward side of the mountain. In the valleys of Switzerland it is regarded as a health hazard, producing migraine and high blood pressure. A similar wind, the chinook, is found on the eastern slopes of the Rocky Mountains in North America.

fold a bend in ◊beds or layers of rock. Folds are caused by pressures within the Earth's crust resulting from plate-tectonic activity. They play an important role in landscape formation and can be eroded to form ◊escarpments, giving rise to an undulating ◊topography.

food chain in ecology, a sequence showing the feeding relationships between organisms in a particular ◊ecosystem. Each organism depends on the next lowest member of the chain for its food. Since many organisms feed at several different levels (for example, some animals feed on both vegetable matter and meat), the relationships often form a complex web rather than a simple chain.

food processing treating a foodstuff to make it more palatable or digestible, or to preserve it from spoilage. Traditional forms of processing include flour-milling, bread-making, yoghurt- and cheese-making, brewing, and various methods of ***food preservation***, such as salting, smoking, pickling, drying, bottling, and preserving in sugar. Modern food technology still employs traditional methods but also uses many new processes, such as refrigeration, freeze-drying, canning, and irradiation, which allow a wider range of foodstuffs to be preserved. Chemical additives, such as flavourings, preservatives, antioxidants, emulsifiers, and colourings, may also be introduced during processing. These methods have enabled farmers and food manufacturers to transport their products to more distant markets.

footloose industry industry that can be sited in any of a number of places, often because transport costs are unimportant. Such industries may have raw materials that are commonly available; for example, a

food chain

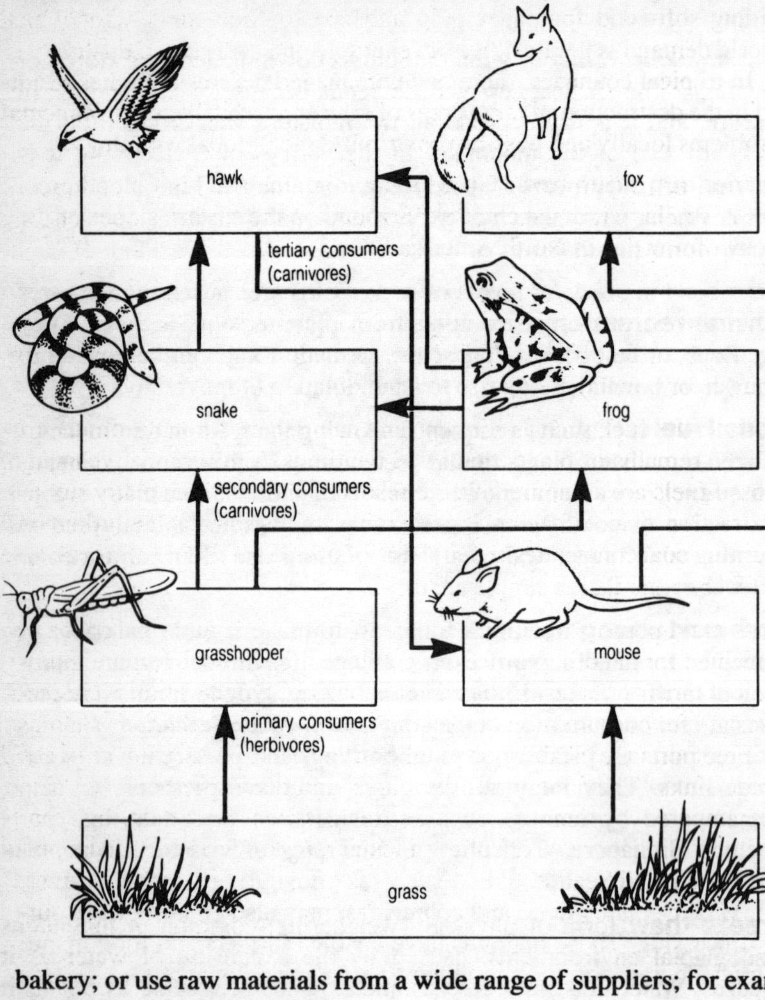

bakery; or use raw materials from a wide range of suppliers; for example, motor-vehicle assembly.

forestry the large-scale management of trees for commercial or recreational and conservation purposes. Forestry has often been confined to the planting of a single species, such as a rapid-growing conifer providing softwood for paper pulp and construction timber, for which world demand is greatest. It is an example of a ◊primary industry.

In tropical countries, rapid and unmanaged deforestation has resulted in the destruction of large areas of rainforest, causing environmental problems locally and possibly contributing to ◊global warming.

formal employment any job where the employee has a contract of employment, pays taxes, and may be provided with a pension scheme. These form the majority of jobs in industrialized (developed) countries.

formal recreation recreation that has to be paid for, have special facilities, or be officially organized in some way; for example, clubs, squash, or bowling.

fossil fuel fuel, such as coal, oil, and natural gas, formed from the fossilized remains of plants that lived hundreds of millions of years ago. Fossil fuels are a ◊nonrenewable resource and will eventually run out. Extraction of coal causes considerable environmental pollution, and burning coal contributes to problems of ◊acid rain and the ◊greenhouse effect.

free port port or sometimes a zone within a port, where cargo may be accepted for handling, processing, and reshipment without the imposition of tariffs or taxes. Duties and tax become payable only if the products are for consumption in the country to which the free port belongs.

Free ports are established to take advantage of a location with good trade links. They facilitate the quick entry and departure of ships, unhampered by lengthy customs regulations. Important free ports include Singapore, Copenhagen, New York, Gdańsk, Macao, San Francisco, and Seattle.

freeze-thaw form of physical ◊weathering, common in mountains and glacial environments, caused by the expansion of water as it freezes. Water in a crack freezes and expands in volume by 9% as it turns to ice. This expansion exerts great pressure on the rock, causing the crack to enlarge. After many cycles of freeze-thaw, rock fragments may break off to form ◊scree slopes.

freeze-thaw

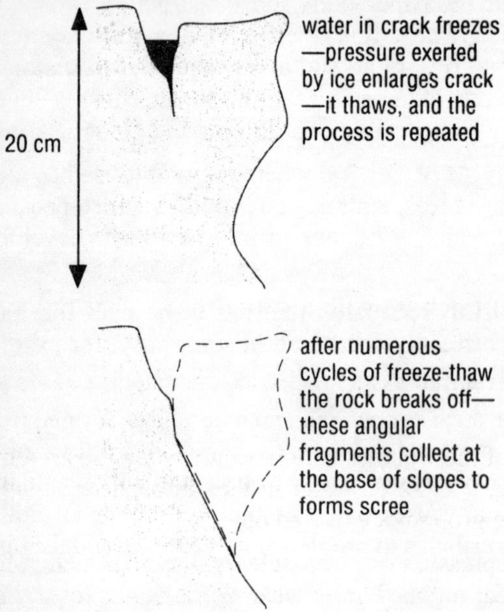

For freeze-thaw to operate effectively the temperature must fluctuate regularly above and below 0°C. It is therefore uncommon in areas of extreme and perpetual cold, such as the polar regions.

front the boundary between two air masses of different temperature or humidity. A ***cold front*** marks the line of advance of a cold air mass from below, as it displaces a warm air mass; a ***warm front*** marks the advance of a warm air mass as it rises up over a cold one. As air rises at a front, it cools and condenses, leading to the formation of cloud and often rain. A warm front brings a long period (about four to six hours) of steady rain whereas a cold front brings a shorter period (about two to three hours) of heavy rain.

Fronts are usually associated with ◊depressions (areas of low atmospheric pressure found in temperate latitudes), which move from west to east in the northern hemisphere. Because it moves more rapidly, a cold

front

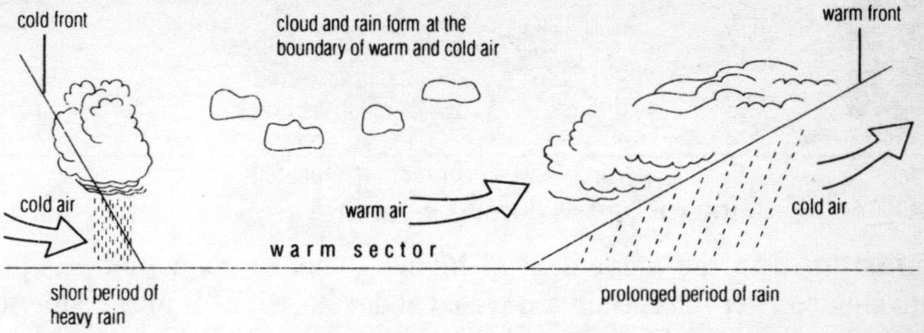

front will tend to catch up with a warm front, forming an ***occluded front***.

frontal rainfall rainfall associated with the meeting of air masses at ◊fronts.

frost condition of the weather which occurs when the air temperature is below freezing, 0°C/32°F. Water in the atmosphere is deposited as ice crystals on the ground or exposed objects.

frost hollow depression or steep-sided valley in which cold air collects on calm, clear nights. Under clear skies, heat is lost rapidly from ground surfaces, causing the air above to cool and flow downhill (as ◊katabatic wind) to collect in valley bottoms. Fog may form under these conditions and, in winter, temperatures may be low enough to cause frost.

frost shattering alternative name for ◊freeze-thaw.

fuelwood wood that is used as a source of energy or fuel, usually in poor countries. Its uncontrolled collection from diminishing areas of common land may be one cause of ◊deforestation.

Fulani member of a W African culture from the southern Sahara and Sahel. The Fulani are traditionally nomadic pastoralists and traders, and practise ◊transhumance. Fulani groups are found in Senegal, Guinea, Mali, Burkina Faso, Niger, Nigeria, Chad, and Cameroon.

G

GDP abbreviation for ◊gross domestic product.

gentrification the movement of higher social or economic groups into an area after it has been renovated and restored. This may result in the outmigration of the people who previously occupied the area. Often the classification of an area as a conservation area encourages gentrification. It is one strategy available to planners in urban renewal schemes within the ◊inner city.

geo narrow coastal inlet often following a line of weakness, such as a ◊fault, in the coastal rock.

geography the study of the Earth's surface: its topography, climate, and physical conditions, and how these factors affect people and society. It is usually divided into ***physical geography***, dealing with landforms and climates, and ***human geography***, dealing with the distribution and activities of peoples on Earth.

geological time time scale embracing the history of the Earth from its physical origin to the present day. Geological time is traditionally divided into eons (Phanerozoic, Proterozoic, and Archaean), which in turn are divided into eras, periods, epochs, and ages.

geology science of the Earth, its origin, composition, structure, and history.

geothermal energy energy extracted for heating and electricity generation from natural steam, hot water, or hot dry rocks in the Earth's crust. Water is pumped down through an injection well and passes through joints in the hot rocks. It then rises to the surface through a recovery well and may be converted to steam or run through a heat exchanger. Dry steam may be directed through turbines to produce electricity.

geothermal energy

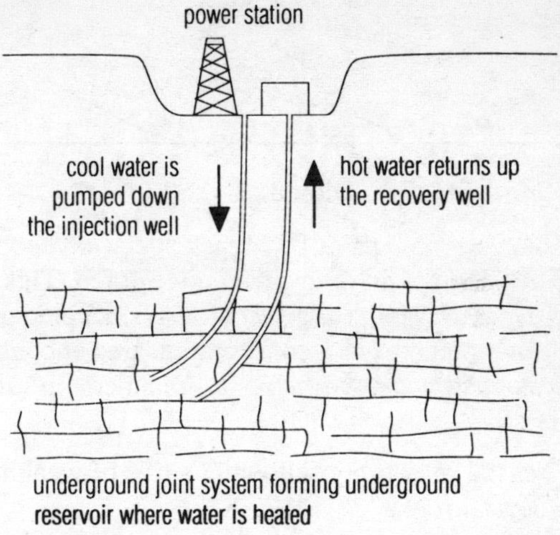

It is an important source of energy in volcanically active areas such as Iceland and New Zealand.

geyser natural spring that intermittently discharges an explosive column of steam and hot water into the air.

One of the most remarkable geysers is Old Faithful, in Yellowstone National Park, Wyoming, USA. Geysers also occur in New Zealand and Iceland.

Giant's Causeway stretch of hexagonal ◊basalt columns forming a headland on the north coast of Antrim, Northern Ireland.

glacial deposition the laying-down of rocky material once carried by a glacier. When ice melts, it deposits the material that it has been carrying. The material dumped on the valley floor forms a deposit called till or boulder clay. It comprises angular particles of all sizes from boulders to clay. Till can be moulded by ice to form drumlins, egg-shaped hills. At the snout of the glacier, material piles up to form a ridge called a terminal moraine. Small depositional landforms may also result from glacial deposition, such as kames (small mounds) and kettle holes (small depressions, often filled with water).

glacial deposition

features associated with glacial deposition

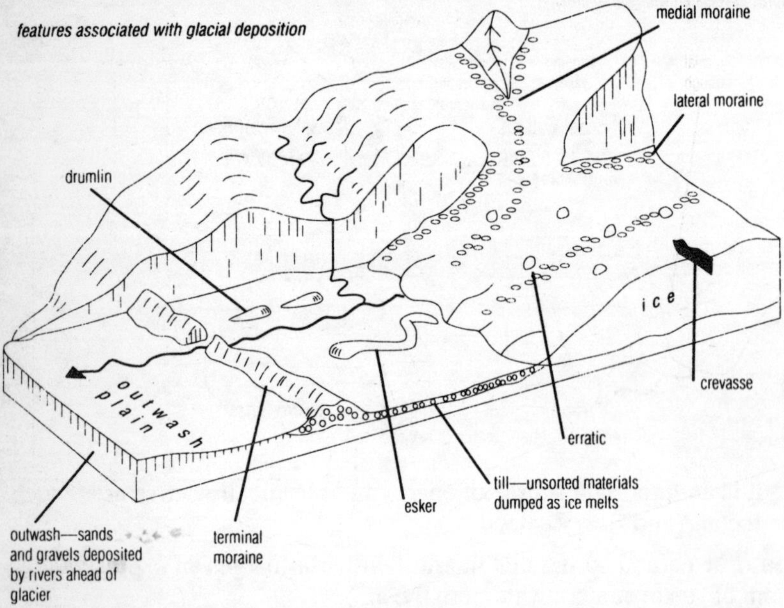

Meltwater flowing away from a glacier will carry some of the till many kilometres away. This sediment will become rounded (by the water) and, when deposited, will form a gently sloping area called an outwash plain. Several landforms owe their existence to meltwater – these are called ***fluvioglacial landforms*** and include the long ridges called eskers. Meltwater may fill depressions eroded by the ice to form ribbon lakes.

In the UK, features of glacial deposition can be seen in East Anglia (till) and Dumfries (drumlins).

glacial erosion the wearing-down and removal of rocks and soil by a glacier. Glacial erosion forms impressive landscape features, including the glacial trough (valley), arêtes (steep ridges), corries (enlarged hollows), and pyramidal peaks (high mountain peaks).

glacial erosion

features associated with glacial erosion

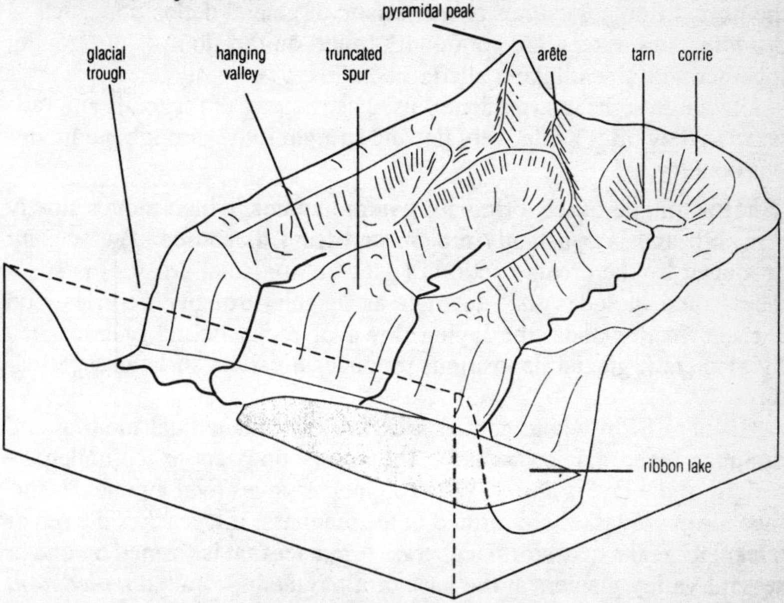

Ice is a powerful agent of erosion. It can bulldoze its way down a valley, eroding away spurs to form truncated spurs. Rock fragments below the ice will abrade the valley floor, leading to overdeepening. Loose fragments will be plucked away from the bedrock to form jagged surfaces such as those on a ◊roche moutonnée. Glacial erosion is made more effective by the process of freeze-thaw (a form of physical weathering), which weakens a rock surface and also provides material for abrasion.

In the UK, features of glacial erosion can be seen in Snowdonia, the Lake District, and the Highlands of Scotland.

glacial trough or ***U-shaped valley*** steep-sided, flat-bottomed valley formed by a glacier. The erosive action of the glacier and of the debris carried by it results in the formation not only of the trough itself but also of a number of associated features, such as truncated spurs (projections of rock that have been sheared off by the ice) and hanging valleys (smaller glacial valleys that enter the trough at a higher level than the trough floor). Features characteristic of glacial deposition, such as drumlins and eskers, are commonly found on the floor of the trough, together with linear lakes called ribbon lakes.

Glacial troughs are frequently used as routeways for roads and railways as they offer a relatively flat and straight route through mountainous country.

glacier tongue of ice, originating in mountains, which moves slowly downhill and is constantly replenished from its source. The scenery produced by the erosive action of glaciers (◊glacial erosion) is spectacular and includes such features as ◊glacial troughs, ◊corries, and ◊arêtes. In lowlands, the laying down of rocky debris once carried by glaciers (◊glacial deposition) produces a variety of landscape features.

Glaciers form where annual snowfall exceeds annual melting and drainage (see ◊glacier budget). The snow compacts to ice under the weight of the layers above. When a glacier moves over an uneven surface, deep crevasses are formed in the ice mass; if it reaches the sea or a lake, it breaks up to form icebergs. A glacier that is formed by one or several valley glaciers at the base of a mountain is called a ***piedmont*** glacier. A body of ice that covers a large land surface or continent – for example, Greenland or Antarctica – and flows outward in all directions is called an ***ice sheet***.

glacier budget in a glacier, the balance between ◊accumulation (the addition of snow and ice to the glacier) and ◊ablation (the loss of snow and ice by melting and evaporation). If accumulation exceeds ablation the glacier will advance; if ablation exceeds accumulation it will probably retreat.

The rate of advance and retreat of a glacier is usually only a few centimetres a year.

GNP abbreviation for ◊gross national product.

Gondwanaland or ***Gondwana*** southern land mass formed 200 million years ago by the splitting of the single world continent ◊Pangaea. (The northern land mass was ◊Laurasia.) It later fragmented into the continents of South America, Africa, Australia, and Antarctica, which then drifted slowly to their present positions.

good in economics, any product. Goods may be divided into a hierarchy of high-order and low-order goods. High-order goods include expensive ◊comparison goods, such as washing machines, furniture, and video cassette recorders, which the shopper will buy only after comparing different models. Low-order goods are usually cheap ◊convenience goods, such as groceries, which the shopper buys frequently and locally.

◊Desire lines may be drawn on maps to indicate the distance that people are prepared to travel in order to obtain a particular good; the longer the desire line, the higher the order of that good.

gorge narrow steep-sided valley (or canyon) that may or may not have a river at the bottom. A gorge may be formed as a ◊waterfall retreats upstream, eroding away the rock at the base of a river valley; or it may be caused by ◊rejuvenation, when a river begins to cut downwards into its channel once again (for example, in response to a fall in sea level). Gorges are common in limestone country, where they are commonly formed by the collapse of the roofs of underground caverns.

Examples of gorges in the UK are Winnats Pass in Derbyshire and the Avon Gorge in Bristol.

gradient the slope of a piece of land or a line on a graph. The gradient shows the vertical rise or fall over a certain horizontal distance. Gradient may be expressed in simple terms – flat, gentle, or steep – or numerically as a percentage or fraction; for example, 20% or 1/5.

granite coarse-grained ◊igneous rock, typically consisting of the minerals quartz, feldspar, and mica. It may be pink or grey, depending on the composition of the feldspar. Granites are chiefly used as building materials.

Granites often form large intrusions in the core of mountain ranges, and they are usually surrounded by zones of metamorphosed rock (see ◊metamorphic rock). Granite areas have characteristic moorland scenery. In exposed areas the bedrock may be weathered along joints and cracks to produce a ◊tor, consisting of rounded blocks that appear to have been stacked upon one another.

graph pictorial representation of statistical data (for example, a ◊pie chart) or of the mathematical relationship between two or more variables (for example, a ◊histogram and a ◊scatter diagram).

gravel coarse ◊sediment consisting of pebbles or small fragments of rock, originating in the beds of lakes and streams or on beaches. Gravel is quarried for use in road building, railway ballast, and for an aggregate in concrete. It is obtained from quarries known as gravel pits, where it is often found mixed with sand or clay. Some gravel deposits also contain ◊placer deposits of metal ores (chiefly tin) or free metals (such as gold and silver).

green belt area surrounding a large city, officially designated not to be built upon but preserved where possible as open space (for agricultural and recreational use). In the UK the first green belts were introduced in 1947 around conurbations such as London in order to prevent ◊urban sprawl. ◊New towns were set up to take the overspill population.

greenfield site land that is a potential ◊industrial location but has not yet been used for urban or industrial development. These sites are attractive to commercial development because site-clearance costs are saved, and the environmental quality of such land may be high.

Many high-tech industries occupy former greenfield sites close to motorway junctions, often at the edge of urban areas.

greenhouse effect phenomenon of the Earth's atmosphere by which solar radiation, absorbed by the Earth and re-emitted from its surface, is prevented from leaving by various gases in the air. The result is a rise in the Earth's temperature. The main greenhouse gases are carbon dioxide, methane, and chlorofluorocarbons (CFCs). Fossil-fuel consumption and forest fires are the main causes of carbon dioxide buildup; methane is a byproduct of agriculture (rice, cattle, sheep). Water vapour is another greenhouse gas.

The United Nations Environment Programme estimates that by 2025 average world temperatures will have risen by 1.5°C with a consequent rise of 20 cm in sea level. Low-lying areas and entire countries would be threatened by flooding and crops would be affected by the change in climate. However, predictions about global warming and its possible climatic effects are tentative and often conflict with each other.

green revolution the change in methods of arable farming introduced in the ◊developing world in the 1940s and 1950s (but abandoned by some countries in the 1980s). The intent is to provide more and better food, albeit with a heavy reliance on chemicals and machinery. Much of the food produced is exported as ◊cash crops, which means that the local diet does not always improve.

Measures include the increased use of tractors and other machines, artificial fertilizers and pesticides, as well as the breeding of new strains of crop plants (mainly rice, wheat, and corn) and farm animals. In places, the introduction of land reform schemes have evened out land ownership. Much of the work is coordinated by the United Nation's Food and Agriculture Organization (FAO).

The green revolution was initially successful in SE Asia; India doubled its wheat yield in 15 years, and the rice yield in the Philippines rose by 75%. However, yields have levelled off in many areas and some countries that cannot afford the dams, fertilizers, and machinery required, have adopted ◊appropriate, or intermediate, technologies. High-yield varieties of cereal plants require large amounts of nitrate fertilizers per hectare, more than is available to small farmers in poor countries. The rich farmers therefore enjoy bigger harvests, and the gap between rich and poor has grown even wider.

grid the network by which electricity is generated and distributed over a region or country; see ◊national grid.

grid reference on a map, numbers that are used to show location. The numbers at the bottom of the map (eastings) are given before those at the side (northings). On British Ordnance Survey maps, a four-figure grid reference indicates a specific square, whereas a six-figure grid reference indicates a point within a square.

gross domestic product (GDP) the total value of all the goods and services produced within a country each year. It is equivalent to gross national product (GNP) less income from investments abroad but including the production of foreign-owned firms within the country.

gross national product (GNP) the total value of all the goods and services produced annually by the firms owned by a nation. It includes income from investments abroad but excludes income from foreign-owned firms within the country. GNP is used to indicate the level of economic wealth of a country or as a measurement of development. It is often expressed in US dollars.

ground water water collected underground in porous rock strata; it emerges at the surface as springs and streams. The ground water's upper level is called the ◊water table. Beds of rock that are filled with ground water are called ***aquifers***. Recent research suggests that usable ground water amounts to more than 90% of all the fresh water on Earth; however, keeping such supplies free of pollutants is of critical environmental concern.

The force of gravity makes water run 'downhill' underground just as it does above the surface. The greater the slope and the permeability, the greater the speed. However, the speed of movement is very slow compared with other forms of water transfer (see ◊hydrological cycle).

growth pole point within an area where economic growth is concentrated. This growth may encourage further development in the surrounding area (through the ◊multiplier effect), especially in areas of industrial decline or stagnation, such as ◊assisted areas. Growth poles are often towns (like Brasília in Brazil) or parts of towns that are gaining industry and expanding. They may be set up as part of a regional planning policy to help development in certain areas; for example, the Mezzogiorno, a region of poor economic performance in S Italy.

groyne wooden or concrete barrier built at right angles to a beach in order to block the movement of material along the beach by ◊longshore drift. Groynes are usually successful in protecting individual beaches, but because they prevent beach material from passing along the coast they can mean that other beaches, starved of sand and shingle, are in danger of being eroded away by the waves. This happened, for example, at Barton-on-Sea in Hampshire, England, in the 1970s, following the construction of a large groyne at Bournemouth.

gryke enlarged ◊joint that separates blocks of limestone (clints) in a ◊limestone pavement.

Gulf Stream warm ◊ocean current that flows north from the warm waters of Gulf of Mexico. Part of the current is diverted east across the Atlantic, where it is known as the ***North Atlantic Drift***, and warms what would otherwise be a colder climate in the British Isles and NW Europe.

H

hail precipitation in the form of pellets of ice (hailstones). It is caused by the circulation of moisture in strong ◊convection currents, usually associated with thunderstorms.

hamlet small rural settlement that is more than just an isolated dwelling but not large enough to be a village. Typically it has 11–100 people, half a dozen houses, and, in the UK, maybe a church and a pub.

hanging valley valley that joins a larger glacial trough at a higher level than the trough floor. During glaciation the ice in the smaller valley was unable to erode as deeply as the ice in the trough, and so the valley was left perched high on the side of the trough when the ice retreated. A river or stream flowing along the hanging valley often forms a waterfall as it enters the trough.

hay preserved grass used for winter livestock feed. The grass is cut and allowed to dry in the field before being removed for storage in a barn.

The optimum period for cutting is when the grass has just come into flower and contains most feed value. During the natural drying process, the moisture content is reduced from 70-80% down to a safe level of 20%. In normal weather conditions, this takes from two to five days during which time the hay is turned by machine to ensure even drying. Hay is normally baled before removal from the field.

headward erosion the backwards erosion of material at the source of a river or stream. Broken rock and soil at the source are carried away by the river, causing erosion to take place in the opposite direction to the river's flow. The resulting lowering of the land behind the source will, over time, cause the river to flow backwards – possibly into a neighbouring valley to capture another river (see ◊river capture).

heat island large town or city that is warmer than the surrounding countryside. The difference in temperature is most pronounced during the winter, when the heat given off by the city's houses, offices, factories, and vehicles raises the temperature of the air by a few degrees.

heavy industry industry that processes large amounts of bulky raw materials. Examples are the iron and steel industry, shipbuilding, and aluminium smelting. Heavy industries are often tied to locations close to their supplies of raw materials.

hedge or ***hedgerow*** row of closely planted shrubs or low trees, generally acting as a land division and windbreak. Hedges also serve as a source of food and as a refuge for wildlife. Between 1945 and 1985, 25% of Britain's hedgerows were destroyed (a length that would stretch seven times around the equator) in order to accommodate altered farming practices and larger machinery. Hedges are further threatened by bad management and spray drift of pesticides.

herbicide or ***weedkiller*** any chemical used to destroy plants or check their growth.

hierarchy order of rank or importance. Many geographical features are arranged in this way; for example, settlements, services, and rivers. Important (high-order) features within a hierarchy occur less frequently than lower-order features; for example, there are fewer cities than villages in a country.

high order (of a settlement, service, or good) at the top of its ◊hierarchy. High-order features are infrequently provided and require large ◊threshold populations to sustain them. They may have an extensive ◊range and therefore a large ◊sphere of influence.

high-tech industry any industry that makes use of advanced technology. The largest high-tech group is the fast-growing electronics industry and especially the manufacture of computers, microchips, and telecommunications equipment.

The products of these industries have low bulk but high value, as do their components. Silicon Valley in the USA and Silicon Glen in Scotland are two areas with high concentrations of such firms. High-tech industries may prefer ◊greenfield sites.

high-yield variety crop that has been specially bred or selected to produce more than the natural varieties of the same species. During the

histogram

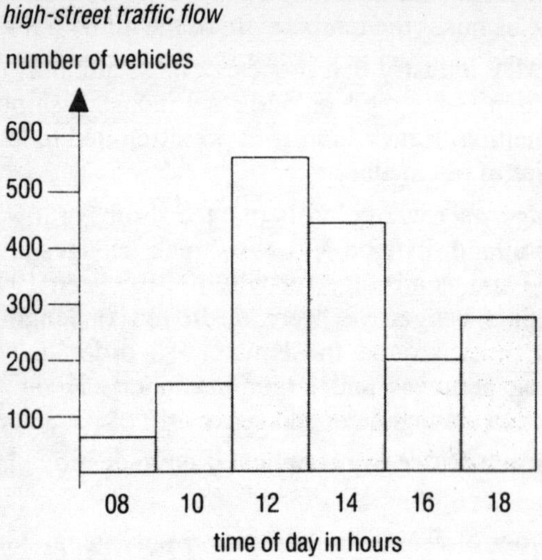

1950s and 1960s, new strains of wheat and maize were developed to reduce the food shortages in poor countries (the ◊green revolution). Later, IR8, a new variety of rice that increased yields by up to six times, was developed in the Philippines. Strains of crops resistant to drought and disease were also developed. High-yield varieties require large amounts of expensive artificial fertilizers and sometimes pesticides for best results.

hinterland area that is served by a port or settlement (the ◊central place) and included in its ◊sphere of influence. The city of Rotterdam, the Netherlands, is the hinterland of a port.

histogram graph in which the data are shown as bars or columns. The height of these depends on the frequency of the values (or groups of values) of the data. Another way of presenting data is the ◊pie chart.

historical inertia another term for ◊industrial inertia.

honeypot site area that is of special interest or appeal to tourists. At peak times, honeypot sites may become crowded and congested, and noise and litter may eventually spoil such areas.

Some developing countries, such as Kenya, have encouraged the development of honeypot sites in order to earn extra income. These areas often have facilities not found elsewhere in the country. In the UK, Dartmoor National Park is an example of a honeypot site.

hops the female fruit-heads of the hop plant *Humulus lupulus*, used to flavour beer. In the UK, hops are grown in Kent, Hereford, and Worcester.

horticulture the growing of flowers, fruit, and vegetables. Horticulture is practised in gardens and orchards, and in millions of acres of land devoted to vegetable farming. Some areas, like California, have specialized in horticulture because they have the mild climate and light fertile soil most suited to these crops.

In Britain, over 200,000 hectares are devoted to commercial horticulture; vegetables account for almost three-quarters of the produce.

hot spot localized thinning of the Earth's ◊crust where molten rock, or magma, escapes to the surface to form a volcano. For example, the Hawaiian Islands in the Pacific Ocean. Hot spots should not be confused with areas of volcanic activity at plate margins (see ◊plate tectonics).

humidity the quantity of water vapour in a given volume of the atmosphere (absolute humidity), or the ratio of the amount of water vapour in the atmosphere to the saturation value at the same temperature (relative humidity). At ◊dew point the relative humidity is 100% and the air is said to be saturated. Condensation (the conversion of vapour to liquid) may then occur. Relative humidity is measured by various types of ◊hygrometer.

humus component of ◊soil consisting of decomposed or partly decomposed organic matter, dark in colour and usually richer towards the surface. It is an important source of minerals in soil fertility.

hurricane or ***tropical cyclone*** or ***typhoon*** at latitudes, an intense ◊depression (region of very low atmospheric pressure). Hurricanes originate between 5° and 20° north or south of the equator when the surface temperature of the ocean is above 27°C. A central calm area, called the ***eye***, is surrounded by inwardly spiralling winds (anticlockwise in the northern hemisphere) of up to 320 kph. A hurricane is accompanied by lightning and torrential rain, and can cause extensive damage. On the ◊Beaufort scale, it is a wind of force 12 or more.

The most intense hurricane recorded in the Caribbean/Atlantic sector was Hurricane Gilbert in 1988, with sustained winds of 280 kph and gusts of over 320 kph. In Oct 1987 and Jan 1990, winds of near-hurricane strength were experienced in S England. Although not technically hurricanes, they were the strongest winds there for three centuries.

hydration form of ◊chemical weathering caused by the expansion of certain minerals as they absorb water. The expansion weakens the parent rock and may cause it to break up.

hydraulic action erosive force exerted by water (as distinct from the forces exerted by rocky particles carried by water). It can wear away the banks of a river, particularly at the outer curve of a meander (bend in the river), where the current flows most strongly.

Hydraulic action occurs as a river tumbles over a waterfall to crash onto the rocks below. It will lead to the formation of a plunge pool below the waterfall. The hydraulic action of ocean waves and turbulent currents forces air into rock cracks, and therefore brings about erosion by ◊cavitation.

hydraulic radius measure of a river's ◊channel efficiency (its ability to discharge water), used by water engineers to assess the likelihood of flooding.

The hydraulic radius of a channel is defined as the ratio of its cross-sectional area to its wetted perimeter (the part of the cross-section that is in contact with the water).

In equation terms hydraulic radius = channel's cross-sectional area/wetted perimeter.

The greater the hydraulic radius, the greater the efficiency of the channel and the less likely the river is to flood. The highest values occur when channels are deep, narrow, and semi-circular in shape.

hydroelectric power (HEP) electricity generated by moving water. In a typical HEP scheme water stored in a reservoir, often created by damming a river, is piped into water turbines, coupled to electricity generators. In ◊pumped storage plants, water flowing through the turbines is recycled. A ◊tidal power station exploits the rise and fall of the tides. About one-fifth of the world's electricity comes from HEP.

HEP is an example of a ◊renewable resource. In the UK it provides 1.3% of total electricity production; stations are located in the mountainous areas of Scotland and North Wales where there is fast-flowing water.

hydrograph graph showing how the discharge of a river varies with time. By studying hydrographs, water engineers can predict when flooding is likely and take action to prevent its taking place.

A hydrograph shows the time lag, or delay, between the occurrence of a rainstorm and the resultant rise in discharge, and the length of time taken for that discharge to peak. The shorter the time lag and the higher the peak, the more likely it is that flooding will occur. Factors likely to give short time lags and high peaks include heavy rainstorms, steep slopes, deforestation, poor soil quality, and the covering of surfaces with impermeable substances such as tarmac and concrete. Actions taken by water engineers to increase time lags and lower peaks include planting trees in the drainage basin of a river.

hydrograph

a flood hydrograph

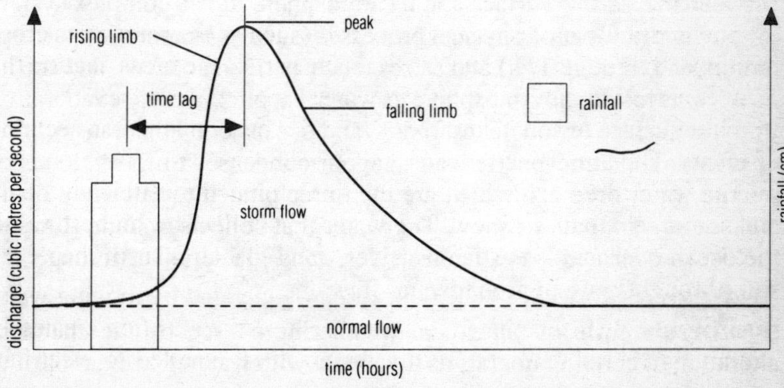

hydrological cycle

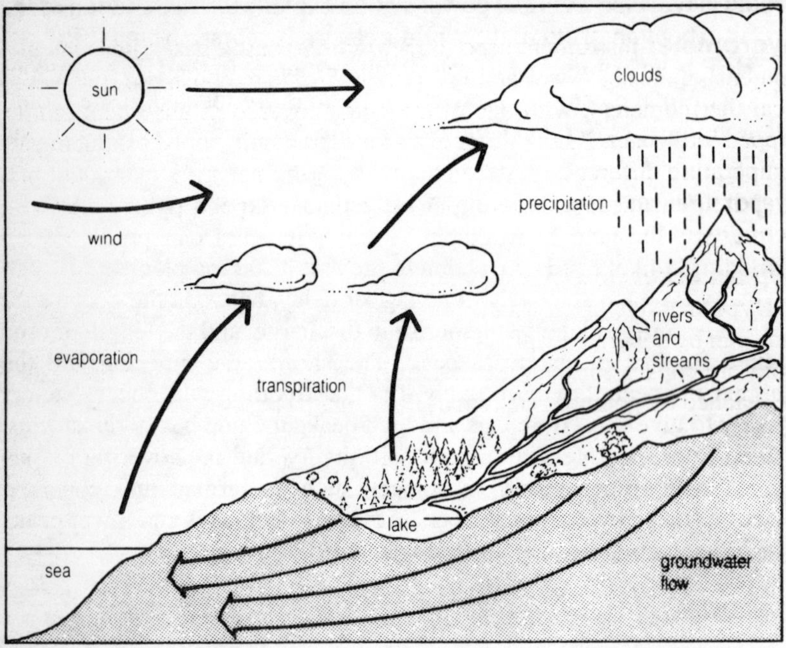

hydrological cycle or ***water cycle*** the cycle by which water moves between the Earth's surface and its atmosphere. It is a complex system involving a number of physical processes (such as evaporation, precipitation, and throughflow) and stores (such as rivers, oceans, and soil).

Water is lost to the atmosphere as water vapour either by evaporation from the surface of soil, lakes, rivers, and oceans or by the transpiration of plants. The atmospheric water vapour condenses to form clouds of minute water droplets, which are then precipitated to fall on the land and sea as rain, hail, or snow. The water that collects on land flows to the ocean overland—as streams, rivers, and glaciers—or through the soil (throughflow and groundwater flow).

hydrolysis form of ◊chemical weathering caused by the chemical alteration of certain minerals as they react with water. For example, the mineral feldspar in granite reacts with water to from a white clay called ◊china clay.

hygrometer instrument used to measure the humidity (water vapour content) of the air. A wet-and-dry-bulb hygrometer consists of two vertical thermometers, with one of the bulbs covered in absorbent cloth dipped into water. As the water evaporates, the bulb cools, producing a temperature difference between the two thermometers. The amount of evaporation, and hence cooling of the wet bulb, depends on the relative humidity of the air.

hypermarket very large ◊supermarket.

Hypermarkets are normally built on the outskirts of towns where land can be bought cheaply and there is ample space for parking. Locations are chosen for their ◊accessibility (the ease with which they can be reached by potential customers).

I

ice age any period in the Earth's history in which polar icesheets have spread to cover regions that are not normally covered with ice.

The most recent ice age (the Ice Age) occurred from 2 million to 10,000 years ago. During this period ice advanced southwards from the Arctic and retreated several times. At its maximum it extended over much of Northern Europe and Northern America. In the UK it advanced as far south as a line between the Severn and the Thames estuary. The area to the south of this line was affected by ◊periglacial conditions.

iceberg floating mass of ice, about 80% of which is submerged, rising sometimes to 100 m above sea level. Glaciers that reach the coast become extended into a broad foot; as this enters the sea, masses break off and drift towards temperate latitudes, becoming a danger to shipping.

ice cap body of ice that is larger than a glacier but smaller than an ice sheet. Such ice masses cover mountain ranges, such as the Alps, or small islands. Glaciers often originate from ice caps.

ice sheet body of ice that covers a large land mass or continent; it is larger than an ice cap. During the last ◊ice age ice sheets spead over large parts of Europe and North America. Today there are two ice sheets, covering much of Antarctica and Greenland. About 96% of all present-day ice is in the form of ice sheets.

igneous rock rock formed from cooling magma (molten rock originating in the Earth's mantle).

Igneous rocks that crystallize below the Earth's surface are called ***intrusive*** – they have large crystals produced by slow cooling. Those formed at the surface are called ***extrusive***, or ***volcanic*** – rapid cooling results in small crystals. Granite is an example of an intrusive rock; basalt is an example of an extrusive rock.

immigration and emigration movement of people from one country to another. Immigration is movement to a country; emigration is movement from a country. Immigration or emigration on a large scale are often for economic reasons or because of religious, political, or social persecution (◊push factors). The USA has received immigrants on a larger scale than any other country, more than 50 million during its history.

impermeable rock rock that does not allow water to pass through it – for example, clay, shale, and slate. Unlike ◊permeable rocks, which absorb water, impermeable rocks can support rivers. They therefore experience considerable erosion (unless, like slate, they are very hard) and commonly form lowland areas.

import product or service that one country purchases from another. Imports may be visible (goods) or invisible (services). In the UK, the most significant visible imports are food, beverages, and tobacco, basic materials, and manufactured goods. The most significant invisible imports are travel, tourism, and transport services.

incised meander in a river, a deep steep-sided meander (bend) formed by the severe downwards erosion of an existing meander. Such erosion is usually brought about by the ◊rejuvenation of a river (for example, in response to a fall in sea level).

There are several incised meanders along the middle course of the river Wye, near Chepstow, Gwent.

industrial estate area planned for industry, where space is available for large buildings and further expansion. Industrial estates often have good internal road layouts and occupy accessible sites near road or motorway junctions but away from the central business district. ◊Business parks and ◊science parks are other planned areas of ◊industrial location.

In the UK, the first industrial estates were built in the 1960s.

industrial inertia the tendency of an industry or group of industries to remain in one place even though its original locational needs have changed. For example, in Sheffield, iron and steel are still produced, even though the local supplies of iron ore ran out some time ago. The cost of moving is one reason why industries may stay in a particular place.

industrialization process by which an increasing proportion of a country's economic activity is involved in industry. It is essential for economic development and is largely responsible for the growth of cities (see ◊urbanization).

industrial location the place where industry is found. Traditionally, industries have located near to their markets, raw materials, or sources of labour. A number of factors influence the location of an industry or group of industries, but, generally, industries choose sites where they can make the most profit. Among their options are ◊greenfield sites, ◊industrial estates, and ◊footloose industry.

industrial location

factors affecting industrial location

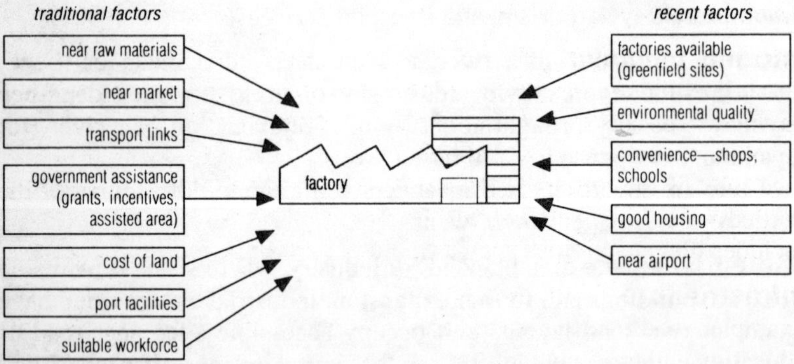

industrial sector any of the different groups into which industries may be divided: primary, secondary, tertiary, and quaternary. ***Primary*** industries extract or use raw materials; for example, mining and agriculture. ***Secondary*** industries are manufacturing industries, where raw materials are processed or components are assembled. ***Tertiary*** industries supply services such as retailing. The ***quaternary*** sector of industry is concerned with the professions and those services that require a high level of skill, expertise, and specialization. It includes education, research and development, administration, and financial services such as accountancy.

infant mortality rate measure of the number of infants dying under one year of age, usually expressed as the number of deaths per 1,000 live births. Improved sanitation, nutrition, and medical care have considerably lowered figures throughout much of the world; for example in the 18th century in the USA and UK infant mortality was about 500 per thousand, compared with under 10 per thousand in 1989. In much of the developing world, however, the infant mortality rate remains high.

In Bangladesh, for example, the infant mortality rate in 1989 was 118 per 1,000.

infiltration the passage of water into the soil. The rate of absorption of surface water by soil (the infiltration capacity) depends on the amount of surface water, the permeability and compactness of the soil, and the extent to which it is already saturated with water. Once in the soil, water may pass into the bedrock to form ◊ground water.

informal employment paid work on a casual basis. Jobs are irregular and people are often self-employed without pensions and without paying taxes. This sort of employment is common in the urban areas of developing countries; for example, in Mexico City. It may involve service jobs of the tertiary ◊industrial sector – shoe cleaning or selling bottled water – as well as craft industries. Informal employment also includes illegal activities such as theft, prostitution, and selling drugs.

infrastructure facilities that service an industrial economy – for example, roads, railways, telephones, energy and water supply, and education and training facilities.

inner city the area that immediately borders the central business district of a town or city. In many cities this is one of the older parts and may suffer from decay and neglect, leading to social problems.

Some theories of urban land use call the outer part of the inner city the 'twilight zone' or zone of transition. There may be poor-quality terraced housing with old manufacturing industry nearby. Low rents and land prices in the inner city may attract newly arriving ethnic groups, who then establish their own sectors within the area. There may be crime and social tension. As industries in the area close down, unemployment rises and land becomes derelict. These so-called brownfield sites are unattractive to new industry and may not attract investment.

In many inner-city areas, such as Glasgow in the 1960s, tenement blocks of housing have been replaced by high-rise blocks of flats as part of urban redevelopment schemes. Other schemes have concentrated on the repair and renovation of existing buildings.

insecticide any chemical pesticide used to kill insects. Among the most effective insecticides are synthetic chlorinated chemicals such as DDT (dichloro-phenyl-trichloroethane). However, these chemicals have proved persistent in the environment and are also poisonous to all animal life, including humans, and are consequently banned in many countries.

insolation the amount of solar radiation (heat energy from the Sun) that reaches the Earth's surface. Insolation varies with season and latitude, being greatest at the equator and least at the poles. At the equator the Sun is consistently high in the sky: its rays strike the equatorial region directly and are therefore more intense. At the poles the tilt of the Earth means that the Sun is low in the sky, and so its rays are slanted and spread out. Winds and ocean currents help to balance out the uneven spread of radiation.

integrated steelworks modern industrial complex where all the steelmaking processes – such as iron smelting and steel shaping – take place on the same site. In the UK, the Redcar/Lackenby works on Teesside is an example of integrated steelworks.

intensive agriculture farming system in which large quantities of inputs, such as labour or fertilizers, are involved over a small area of land. ◊Market gardening is an example. Yields are often much higher than those obtained from ◊extensive agriculture.

interception process by which trees and plants hinder the passage of precipitation (such as rain) on its way to the ground. High rates of interception slow down the transfer of rainwater into rivers and make flooding less likely.

interlocking spur one of a series of spurs (ridges of land) jutting out from alternate sides of a river valley. During glaciation its tip may be sheared off by erosion, creating a ◊truncated spur.

intermediate technology another term for ◊appropriate technology.

International Date Line (IDL) imaginary line that approximately follows the 180° line of longitude. The date is put forward a day when crossing the line going west, and back a day when going east.

intertropical convergence zone (ITCZ) area of heavy rainfall found in the tropics and formed as two air masses converge and rise to form cloud and rain. It moves a few degrees northwards during the northern summer and a few degrees southwards during the southern summer, following the apparent movement of the Sun. The ITCZ is responsible for most of the rain that falls in Africa. The ◊doldrums are also associated with this zone.

intrusion mass of ◊igneous rock that has formed by the 'injection' of molten rock, or ◊magma, into existing cracks beneath the surface of the Earth. By contrast, an extrusion, or volcanic rock mass, is formed at the surface from erupted magma.

Intrusions form a number of landscape features – for example, sills, dykes, laccoliths, and batholiths.

intrusive rock ◊igneous rock formed within the Earth. Magma, or molten rock, cools slowly at these depths to form coarse-grained rocks, such as granite, with large crystals. (◊Extrusive rocks, which are formed on the surface, are usually fine-grained.) A mass of intrusive rock is called an intrusion.

iron ore any mineral from which iron is extracted. The chief iron ores are ***magnetite***, a black oxide, and ***hematite***, or kidney ore, a reddish oxide. Much of the world's iron is extracted in the Commonwealth of Independent States (former USSR). Other important producers are the USA, Australia, France, Brazil, and Canada.

irrigation artificial water supply for dry agricultural areas by means of dams and channels. An example is the channelling of the annual Nile flood in Egypt, which has been done from earliest times to its present control by the Aswan High Dam. Drawbacks to irrigation are that it tends to concentrate salts, ultimately causing soil infertility, and that rich river silt is retained at dams, to the impoverishment of the land and fisheries below them.

island arc curved chain of islands produced by volcanic activity at a ◊destructive margin (where one tectonic plate slides beneath another). Island arcs are common in the Pacific where they ring the ocean on both sides; the Aleutian Islands off Alaska are an example.

isobar line drawn on a map or weather chart linking all places with the same atmospheric pressure (usually measured in millibars). Where the isobars are drawn close together they indicate strong winds and a ◊depression (area of low pressure).

isochrone on a map, a line that joins places that are equal in terms of the time it takes to reach them.

isohyet on a map, a line joining points of equal rainfall.

isoline on a map, a line that joins places of equal value. Examples are contour lines (joining places of equal height), ◊isobars (for equal pressure), isotherms (for equal temperature), and isohyets (for equal rainfall). Isolines are most effective when values change gradually and when there is plenty of data.

isostasy the balance in buoyancy of all parts of the Earth's ◊crust, During an ◊ice age the weight of the ice sheet pushes the continent into the Earth's mantle and once the ice has melted, the continent rises again. The movement causes a relative change in sea-level.

isotherm line on a map linking all places having the same temperature at a given time.

J

joint vertical crack in a rock, formed by compression, usually several metres in length. It differs from a ◊fault in that no displacement has taken place. Joints are common in limestone and granite, and the weathering of joints in these rocks is responsible for the formation of features such as ◊limestone pavements and ◊tors. Joints in coastal rocks are often exploited by the sea to form erosion features such as caves and geos.

K

kame feature, usually in the form of a mound or ridge, formed by the deposition of rocky material carried by a stream of glacial meltwater. Kames are commonly laid down in front of or at the edge of a glacier (kame terrace), and are associated with the disintegration of glaciers at the end of an ice age.

Kames are made of well-sorted rocky material, usually sands and gravels. The rock particles tend to be rounded (by attrition) because they have been transported by water.

kaolin another name for ◊china clay.

karst landscape characterized by remarkable surface and underground forms, created as a result of the action of water on permeable limestone. The feature takes its name from the Karst region on the Adriatic coast of Slovenia, but the name is applied to landscapes throughout the world, the most dramatic of which is found near the city of Guilin in the Guangxi province of China.

katabatic wind cool wind that blows down a valley on calm clear nights. (By contrast, an ◊anabatic wind is warm and moves up a valley in the early morning.) When the sky is clear, heat escapes rapidly from ground surfaces, and the air above the ground becomes chilled. The cold dense air moves downhill, forming a wind that tends to blow most strongly just before dawn.

Cold air blown by a katabatic wind may collect in a depression or valley bottom to create a ◊frost hollow.

Katabatic winds are most likely to occur in the late spring and autumn because of the greater daily temperature differences.

kettle hole pit or depression created by glacial activity. A kettle hole is formed when a block of ice from a receding glacier becomes isolated and buried in glacial debris (till). As the block melts the till

kame

formation of kame features

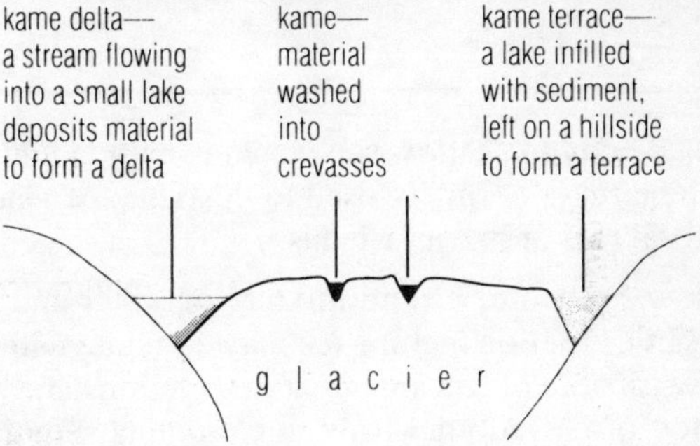

collapses to form a hollow, which may become filled with water to form a kettle lake or pond. Kettle holes range from 5 m to 13 km in diameter, and may exceed 33 m in depth.

knickpoint break in a river slope, often with a waterfall or rapids, resulting from ◊rejuvenation. It marks the junction between the river's old-established long profile and the new profile caused by downcutting.

L

lagoon coastal body of shallow salt water, usually with limited access to the sea. The term is normally used to describe the shallow sea area cut off by a coral reef or barrier islands.

lahar mudflow formed of a fluid mixture of water and volcanic ash. During a volcanic eruption melting ice may combine with ash to form a powerful flow capable of causing great destruction. The lahars created by the eruption of Nevado del Ruiz in Colombia, South America, in 1985 buried 22,000 people in 8 m of mud.

lake body of still water lying in depressed ground without direct communication with the sea. Lakes are common in formerly glaciated regions, along the courses of slow rivers, and in low land near the sea.

Agricultural fertilizers may leach into lakes from the land, causing ◊eutrophication, an excessive enrichment of the water that causes an explosion of aquatic life. This then depletes the lake's oxygen supply until it is no longer able to support life.

land breeze gentle breeze blowing from the land towards the sea and affecting coastal areas. It forms at night in the summer or autumn, and tends to be cool. By contrast, a ◊sea breeze blows from the sea towards the land.

land invasion the movement of people into an area from outside. This may result in changes in the way the land is used. In ◊concentric-ring theory, migrants move into inner-city areas as the original occupants move away towards the outskirts.

land reform policy that the ownership of land should be shared among the peasants and the agricultural workers. In some developing countries, the lack of land reform has meant that the potential benefits of the ◊green revolution have not been spread evenly between rich and poor farmers.

land set-aside scheme policy introduced by the European Community in the late 1980s, as part of the Common Agricultural Policy, to reduce overproduction of certain produce. Farmers are paid not to use land but to keep it ◊fallow. The policy may bring environmental benefits by limiting the amount of fertilizers and pesticides used.

landslide sudden downward movement of a mass of soil or rocks from a cliff or steep slope. Landslides happen when a slope becomes unstable, usually because the base has been undercut or because materials within the mass have become wet and slippery.

A ***mudflow*** happens when soil or loose material is soaked so that it no longer adheres to the slope; it forms a tongue of mud that reaches downhill from a semicircular hollow. A ***slump*** occurs when the material stays together as a large mass, or several smaller masses, and these may form a tilted steplike structure as they slide. A ***landslip*** is formed when ◊beds of rock dipping towards a cliff slide along a lower bed. Earthquakes may precipitate landslides.

lateral moraine linear ridge of rocky debris deposited near the edge of a ◊glacier. Much of the debris is material that has fallen from the valley side onto the glacier's edge, having been weathered by ◊freeze-thaw (the alternate freezing and thawing of ice in cracks); it will, therefore, tend to be angular in nature. Where two glaciers merge, two lateral moraines may join together to form a ***medial moraine*** running along the centre of the merged glacier.

laterite red iron-rich soil characteristic of tropical areas. It may form an impermeable and infertile layer that hinders plant growth.

latifundium (plural ***latifundia***) large-scale ◊farming system common in many Latin American countries, such as Brazil. The land is organized into large, centrally managed estates worked by peasants. These peasants are landless labourers who sell their labour, when conditions permit, for low wages on the estates or commercial plantations.

latitude and longitude imaginary lines used to locate position on the globe. Lines of latitude are drawn parallel to the equator, with 0° at the equator and 90° at the north and south poles. Lines of longitude are

landslide

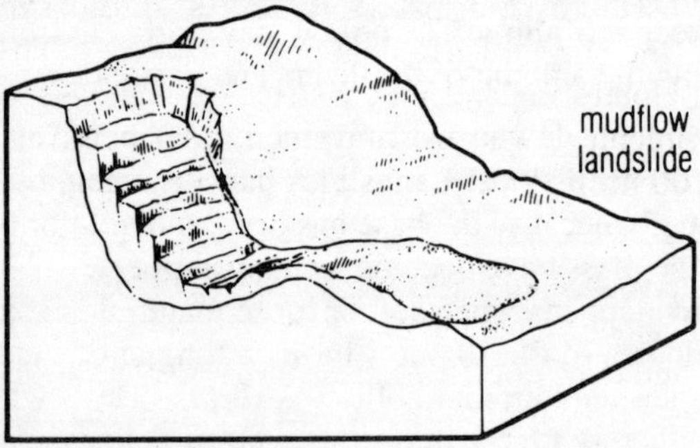

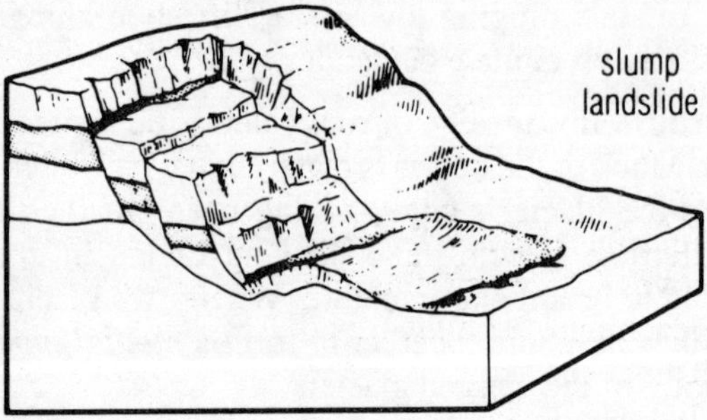

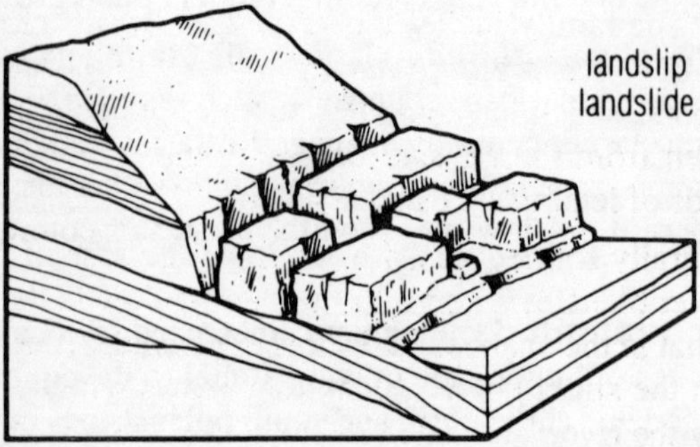

drawn at right angles to these, with 0° (the Prime Meridian) passing through Greenwich, London.

Laurasia northern land mass formed 200 million years ago by the splitting of the single world continent ◊Pangaea. (The southern land mass was ◊Gondwanaland.) It consisted of what was to become North America, Greenland, Europe, and Asia, and is believed to have broken up about 100 million years ago with the separation of North America from Europe.

lava molten rock that erupts from a volcano and cools to form extrusive, or volcanic, ◊igneous rock. Lava that is high in silica is viscous and sticky and does not flow far; it forms a ◊composite volcano. Low-silica lava can flow for long distances and forms a ◊shield volcano.

Lava is different from magma in that it is molten rock on the surface. Magma is molten rock below the surface.

LDC abbreviation for ***less developed country***, another term for any country in the ◊developing world.

leaching process by which substances are washed out of the soil. Fertilizers leached out of the soil find their way into rivers and cause water pollution. In tropical areas, leaching of the soil after the destruction of forests removes scarce nutrients and leads to a dramatic loss of soil fertility. The leaching of soluble minerals in soils can lead to the formation of distinct horizons as different minerals are deposited at successively lower levels.

legume plant of the family Leguminosae, which has a pod containing dry seeds. The family includes peas, beans, lentils, clover, and alfalfa (lucerne). Legumes are important in agriculture because of their specialized roots, which have nodules containing bacteria capable of fixing nitrogen from the air and increasing the fertility of the soil. The edible seeds of legumes are called ***pulses***.

levee naturally formed raised bank along the side of a river channel. When a river overflows its banks, the rate of flow in the flooded area is less than that in the channel, and silt is deposited. After the waters have withdrawn the silt is left as a bank that grows with successive floods. Eventually the river, contained by the levee, may be above the surface of the surrounding flood plain. Notable levees are found on the lower reaches of the Mississippi in the USA and the Po in Italy.

ley area of temporary grassland, sown to produce grazing and hay or silage for a period of one to ten years before being ploughed and cropped. Short-term leys are often incorporated in systems of ◊crop rotation.

life expectancy the average age to which a person, at the time of his or her birth, can expect to live. Life expectancy depends on nutrition, disease control, environmental contaminants, war, stress, and living standards in general.

There is a marked difference between industrialized countries, which generally have a high life expectancy and an ageing population, and the poorest countries, where life expectancy is much shorter. For example, in 1991 life expectancy in the UK was 73 for males and 79 for females, and in Japan was 76 for males and 82 for females. In the same year, life expectancy in Chad was only 39 for males, 41 for females, and in Afghanistan was 43 for males, 44 for females.

In the ◊demographic transition model, life expectancy increases as death rate is reduced.

lignite brown fibrous ◊coal with a relatively low carbon content.

limestone sedimentary rock composed chiefly of calcium carbonate $CaCO_3$, either derived from the shells of marine organisms or precipitated from solution, mostly in the ocean. Various types of limestone are used as building stone.

Limestone is permeable and dissolves gradually in the weak acid of rainwater. It is therefore vulnerable to erosion and weathering by ◊solution, and exhibits a number of characteristic landforms including ◊caves and ◊limestone pavements.

In the UK limestone forms upland areas such as the Pennines and the Mendips, because it does not support surface rivers and therefore erosion is minimal.

limestone pavement bare rock surface resembling a block of chocolate, found on limestone plateaus. It is formed by the weathering of limestone into individual upstanding blocks, called clints, separated

limestone

features of a limestone region

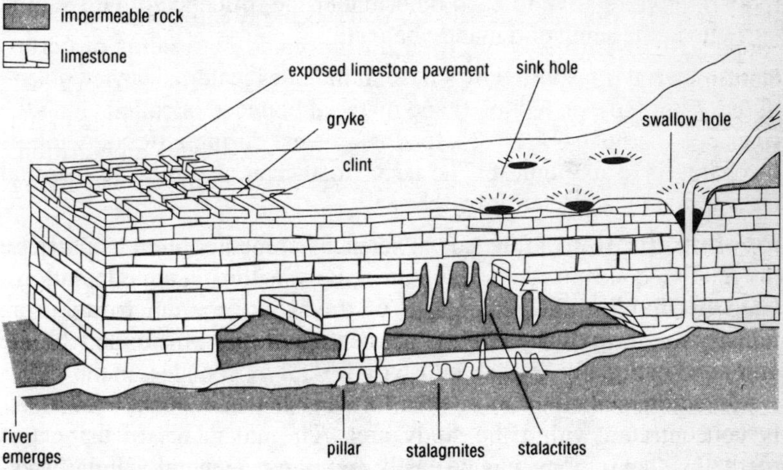

from each other by joints, called grykes. The weathering process is thought to entail a combination of freeze-thaw (the alternate freezing and thawing of ice in cracks) and carbonation (the dissolving of minerals in the limestone by weakly acidic rainwater). Malham Tarn in North Yorkshire is an example of a limestone pavement.

linear development or ***ribbon development*** housing that has grown up along a route such as a road. Many settlements show this ribbon-shaped pattern, since roads offer improved access to the central business district and other areas. Linear development may result in ◊urban sprawl. Other types of settlement pattern are ◊dispersed settlements and ◊nucleated settlements.

liquefaction the conversion of a soft deposit, such as clay, to a jelly-like state by severe shaking. During an earthquake buildings and lines of communication built on materials prone to liquefaction will sink and topple. In the Alaskan earthquake of 1964 liquefaction led to the destruction of much of the city of Anchorage.

lithosphere outermost layer of the Earth's structure, 75 km thick, consisting of the crust and a portion of the upper mantle. It is divided into large rigid areas, or plates, which, according to the ◊plate tectonics model, move relative to each other under the influence of convection currents in the semisolid mantle beneath.

load material transported by a river. It includes material carried on and in the water (suspended load) and material bounced or rolled along the river bed (bedload). A river's load is greatest during a flood, when its discharge is at its highest. The term 'load' can also refer to material transported by a glacier or by the sea.

location quotient in the social sciences, a method used to compare the relative concentration of human activity within an area. It is calculated by dividing the percentage of the workforce employed in an industry in the study area by the percentage of the workforce who are employed nationally in that industry.

A location quotient of more than 1 means that an industry is relatively concentrated within the study area. The advantage of using this method is that regions may be easily compared. Generally, light ◊footloose industries (such as printing) will have similar quotients in most areas, whereas industries that are tied to a raw material (such as iron or steel) or a market will show much greater variations between areas.

locust winged insect similar to a grasshopper. Locusts form large swarms in tropical parts of the world, such as Africa. They can be serious pests, devouring any vegetation or crops in their path, and contributing to the problem of famine. They are partially controlled by spraying concentrated insecticides from aircraft over the insects or the vegetation on which they feed.

loess fertile yellow soil derived from glacial meltwater deposits and accumulated by wind in ◊periglacial regions during the ice ages. There are large deposits of loess in central Europe (Hungary), China, and North America.

longitude see ◊latitude and longitude.

longshore drift the movement of material along a ◊beach. When a wave breaks obliquely, pebbles are carried up the beach in the direction

longshore drift

the action of longshore drift

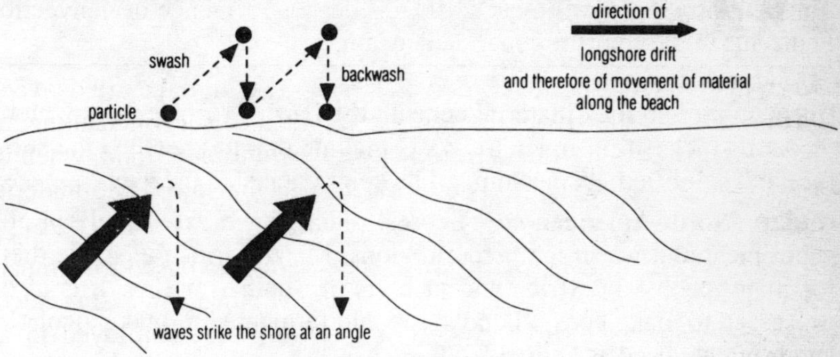

of the wave (swash). The wave draws back at right angles to the beach (backwash), carrying some pebbles with it. In this way, material moves in a zigzag fashion along a beach. Longshore drift is responsible for the formation of ◊spits. Attempts are often made to halt longshore drift by erecting barriers, or ◊groynes, at right angles to the shore.

low order any settlement, service, or good that is at the bottom of its ◊hierarchy. Low order features are frequently provided and require only low ◊threshold populations to sustain them. Their ◊range is limited and their ◊sphere of influence small.

M

magma molten rock material beneath the Earth's surface from which ◊igneous rocks are formed. ◊Lava is magma that has reached the surface and solidified, losing some of its components on the way.

maize (North American ***corn***) cereal plant grown extensively in all subtropical and warm temperate regions. Maize forms the staple diet for many people in Africa and in parts of South America. It is also processed to make corn oil and syrup and fermented to make alcohol, and is widely used as animal feed.

malnutrition condition resulting from a defective diet where certain important food nutrients (such as protein, vitamins, or carbohydrates) are absent. It can lead to deficiency diseases. A related problem is ◊undernourishment.

Malthus theory projection of population growth made by English economist Thomas Malthus in 1793. He based his theory on the ◊population explosion that was already becoming evident in the late 18th century, and argued that the number of people would increase faster than the food supply. Population would eventually reach a resource limit (◊overpopulation). Any further increase would result in a population crash, caused by famine, disease, or war.

Malthus was not optimistic about the outcome and suggested that only 'moral restraint' (birth control) could prevent crisis. Recent famines in Ethiopia and other countries, where drought, civil war, and poverty have reduced agricultural output, might suggest that Malthus was correct. In the UK, however, the agricultural revolution boosted food production, and contraception led to a decline in ◊birth rate; and in some countries ◊population-control policies have been introduced.

mangrove swamp muddy swamp found on tropical coasts and estuaries, characterized by dense thickets of mangrove trees. These low trees are adapted to live in creeks of salt water and send down special breathing roots from their branches to take in oxygen from the air. The roots trap silt and mud, creating a firmer drier environment over time. Mangrove swamps are common in the Amazon delta and along the coasts of W Africa, N Australia, and Florida, USA.

mantle intermediate zone of the ◊Earth between the crust and the core.

manufacturing industry or *secondary industry* industry that involves the processing of raw materials or the assembly of components. Examples are aluminium smelting, car assembly, and computer assembly. In the UK many traditional manufacturing industries, built up in the Industrial Revolution, are now declining in importance; for example, shipbuilding. Developing countries may lack the capital and expertise necessary for these industries.

map diagrammatic representation of an area. Maps are an excellent source of secondary data, providing information on distances, gradients, and landforms within a small area.

Modern maps of the Earth are made using satellites in low orbit to take a series of overlapping photographs. Conventional aerial photography, laser beams, microwaves, and infrared equipment are also used for surveying land. The Ordnance Survey is the official body responsible for the mapping of Britain; it produces maps in a variety of ◊scales, such as the Landranger series (scale 1:50,000). Large-scale maps – for example, 1:25,000 – show greater detail at a local level than small-scale maps; for example, 1:100,000.

Detailed maps requiring constant updating are kept in digital form on computer so that minor revisions can be made without redrafting. There are several different types of mapping technique, including ◊choropleth maps and ◊topological maps.

marble metamorphic rock formed from ◊limestone. It is used in building and sculpture.

marginal land in farming, poor-quality land that is likely to yield a poor return. It is the last land to be brought into production and the first land to be abandoned. Examples are desert fringes in Africa and mountain areas in the UK.

Mariana Trench ocean trench in the NW Pacific off Japan that extends 11,000 m below sea level. It is the lowest region on the Earth's surface.

maritime climate climate typical of coastal areas that is characterized by mild and damp conditions because of the nearness of the sea. Most of the UK has a maritime climate.

market gardening ◊farming system that specializes in the commercial growing of vegetables, fruit, or flowers. It is an ◊intensive agriculture with crops often being grown inside greenhouses on small farms.

Market gardens may be located within easy access of markets, on the fringes of urban areas; for example, in the Home Counties for the London market. Such areas as the Channel Islands, where early crops can be grown outside because of a mild climate, are especially suitable.

market town settlement with a permanent or periodic market (once a week, for instance). This acts as a selling point for goods produced in the surrounding ◊hinterland.

In the UK, a market was often established by means of a royal charter.

marsh low-lying wetland. Freshwater marshes are common wherever groundwater, surface springs, streams, or run-off cause frequent flooding or more or less permanent shallow water. Near the sea, a ◊salt marsh may form on the sheltered side of sand and shingle spits. A marsh is alkaline whereas a ◊bog is acid.

Masai member of an E African people whose territory is divided between Tanzania and Kenya, and who number about 250,000. They were originally warriors and nomads, breeding humped zebu cattle, but some have adopted a more settled life.

mass movement downhill movement of surface materials under their own weight (by gravity). Some types of mass movement are very rapid – for example, ◊landslides, whereas others, such as ◊soil creep, are slow. Water often plays a significant role as it acts as a lubricant, enabling material to move.

mass production manufacture of goods on a large scale, a technique that aims for low unit cost and high output (see ◊economy of scale). In factories mass production is achieved by a variety of means, such as division and specialization of labour and mechanization. These speed up production and allow the manufacture of near-identical, interchangeable parts. Such parts can then be assembled quickly into a finished product on an assembly line.

Many of the machines now used in factories are robots (for example, on car-assembly lines): they work automatically under computer control. Such automation further streamlines production and raises output.

MDC abbreviation for ***more developed country***, another term for a country in the industrialized or ◊developed world.

meander sweeping curve in the course of a river. A meander usually has a steep slope (river cliff) on its outer curve where the velocity of the river, and therefore erosion, is greatest, and a gentle (slip-off) slope on its inner curve where the velocity is slowest. A meander may become so accentuated that it becomes cut off from the course of the river and forms an ◊oxbow lake.

The formation of meanders is still something of a puzzle. Their occurrence is thought to depend upon the gradient (slope) of the land, the nature of the river's discharge, and the type of material being carried. Certainly, meanders are common where the gradient is gentle, the discharge fairly steady (not subject to extremes), and the material carried is fine. Alternating deeps and shallows (called ◊pools and riffles) are also thought to be play a part in the formation of meanders.

mechanical weathering alternative name for ◊physical weathering.

medial moraine linear ridge of rocky debris running along the centre of a glacier. Medial moraines are commonly formed by the joining of two ◊lateral moraines when two glaciers merge.

Mediterranean agriculture ◊farming system found in countries surrounding the Mediterranean Sea. It has developed as a result of the wet autumns and dry summers in this area. Cereal crops are sown in the autumn and harvested in late spring. Trees and vines are grown and crops from them, such as grapes and olives, are collected in the summer after ripening. In addition, goats and sheep are often kept to provide extra income.

Mediterranean climate climate characterized by hot dry summers and warm wet winters. The regions bordering the Mediterranean Sea, California, central Chile, the Cape of Good Hope, and parts of S Australia have such climates.

meltwater water produced by the melting of snow and ice, particularly in glaciated areas. It plays a significant role in the movement of glaciers, as it lubricates the movement of the ice over the ground (without meltwater ice would move very slowly or not at all). Streams of meltwater flowing from glaciers transport rocky materials away from the ice to form ◊outwash. Features formed by the deposition of debris carried by meltwater or by its erosive action are called ***fluvioglacial features***; they include eskers, kames, and outwash plains.

Mercalli scale scale used to measure the effects of an ◊earthquake. It differs from the ◊Richter scale, which measures an earthquake's magnitude.

mesa (Spanish 'table') a flat-topped steep-sided plateau, consisting of horizontal weak layers of rock topped by a resistant formation; in particular, those found in the desert areas of the USA and Mexico. A small mesa is called a butte.

metamorphic rock rock that has been altered by heat or pressure, or both. (If a rock is melted by heat, however, it forms ◊igneous rock upon cooling.) Metamorphic rocks are tough and resistant to erosion, and are often used as building materials.

There are two types of metamorphism. ***Thermal metamorphism*** is brought about by the baking of solid rocks in the vicinity of an igneous intrusion (molten rock, or magma, in a crack in the Earth's crust). It is responsible, for example, for the conversion of limestone to marble. ***Regional metamorphism*** results from the heat and intense pressures associated with the movements and collision of tectonic plates (see ◊plate tectonics). It brings about the conversion of shale to slate, for example.

meteorology scientific observation and study of the atmosphere, so that ◊weather can be accurately forecast. Data from meteorological stations and weather satellites are collated by computer at central agencies such as the Meteorological Office in Bracknell, near London, and weather maps based on current readings are issued at regular intervals. Modern analysis can give useful forecasts for up to six days ahead.

Mercalli scale

value	*description*
I	Only detected by instrument.
II	Felt by people resting.
III	Felt indoors; hanging objects swing; feels like passing traffic.
IV	Feels like passing heavy traffic; standing cars rock; windows, dishes, and doors rattle; wooden frames creak.
V	Felt outdoors; sleepers are woken; liquids spill; doors swing open.
VI	Felt by everybody; people stagger; windows break; trees and bushes rustle; weak plaster cracks.
VII	Difficult to stand upright; noticed by vehicle drivers; plaster, loose bricks, tiles, and chimneys fall; bells ring.
VIII	Car steering affected; some collapse of masonry; chimney stacks and towers fall; branches break from trees; cracks in wet ground.
IX	General panic; serious damage to buildings; underground pipes break; cracks and subsidence in ground.
X	Most buildings destroyed; landslides; water thrown out of canals.
XI	Rails bent; underground pipes totally destroyed.
XII	Damage nearly total; rocks displaced; objects thrown into the air.

metropolitan area another term for ◊conurbation.

microclimate the climate of a small area, such as a woodland, lake, or even a hedgerow. Significant differences can exist between the climates of two neighbouring areas – for example, a town is usually warmer than the surrounding countryside (forming a ◊heat island), and a woodland cooler, darker, and less windy than an area of open land.

Microclimates play a significant role in agriculture and horticulture, as different crops require different growing conditions.

Mid-Atlantic Ridge ◊ocean ridge that runs along the centre of the Atlantic Ocean, for some 14,000 km – almost from the Arctic to the Antarctic. It was formed at a ◊constructive margin between tectonic plates.

The Mid-Atlantic Ridge is central because the ocean crust beneath the Atlantic Ocean has grown outwards from both sides of the ridge at a steady rate during the past 200 million years. Iceland straddles the ridge and was formed by volcanic outpourings.

migration movement of people away from their homes. Migrations may be temporary (for example, holiday makers), seasonal (◊transhumance), or permanent (people moving to cities to find employment); local, national, or international.

The decision to migrate depends on ◊push factors and ◊pull factors between areas, and of the barriers that may exist, such as cost, language, politics, and knowledge. If immigration exceeds emigration, the population of an area will rise beyond the ◊natural increase.

millionaire city or ***million city*** city with more than 1 million inhabitants. In 1985 there were 273 millionaire cities in the world, compared with just two in 1850. Most of these are now found in the developing world, whereas before 1970 most were in industrialized countries.

mineral naturally formed inorganic substance with a particular chemical composition. Minerals are the constituents of ◊rocks. In more general usage, a mineral is any substance economically valuable for mining – including coal and oil, which are of organic origin.

minifundium (plural ***minifundia***) small peasant holding in some South American countries, such as Brazil.

mist low cloud caused by the condensation of water vapour in the lower part of the ◊atmosphere. Mist is less thick than ◊fog, visibility being 1–2 km.

mistral cold, dry, northerly wind that occasionally blows during the winter on the Mediterranean coast of France. It has been known to reach a velocity of 145 kph.

mixed farming farming system where both arable and pastoral farming is carried out. Mixed farming is a lower-risk strategy than ◊monoculture. If climate, pests, or market prices are unfavourable for one crop or type of livestock, another may be more successful and the risk is shared. Animals provide manure for the fields and help to maintain soil fertility.

model simplified version of some aspect of the real word. Models are produced to show the relationships between two or more factors, such as land use and the distance from the centre of a town (for example, ◊concentric-ring theory). Because models are idealized, they give only a general guide to what may happen.

Mohoroviçić discontinuity also ***Moho*** or ***M-discontinuity*** boundary that separates the Earth's crust and mantle, marked by a rapid increase in the speed of earthquake waves.

monoculture farming system where only one crop is grown. In developing countries this is often a ◊cash crop, grown on ◊plantations. Cereal crops in the industrialized world are also frequently grown on a monoculture basis; for example, wheat in the Canadian prairies.

Monoculture allows the farmer to tailor production methods to the requirements of one crop, but it is a high-risk strategy since the crop may fail (because of pests, disease, or bad weather) and world prices for the crop may fall. Monoculture without ◊crop rotation is likely to result in reduced soil quality despite the addition of artificial fertilizers, and it contributes to ◊soil erosion.

monsoon wind pattern that brings seasonally heavy rain to S Asia; it blows towards the sea in winter and towards the land in summer. The monsoon may cause destructive flooding all over India and SE Asia from April to September, leaving thousands of people homeless each year.

moraine rocky debris or ◊till carried along and deposited by a ◊glacier. Material eroded from the side of a glaciated valley and carried along the glacier's edge is called a ***lateral moraine***; that worn from the valley floor and carried along the base of the glacier is called a ***ground moraine***. Rubble dropped at the foot of a melting glacier is called a ***terminal moraine***.

multiple-nuclei theory

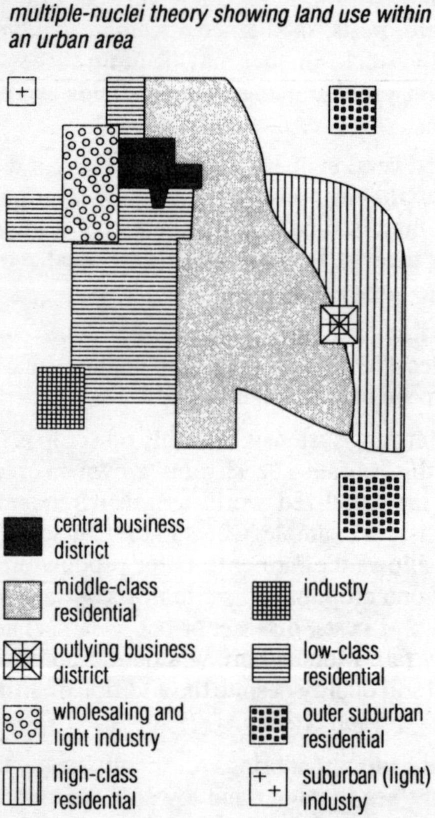

When two glaciers converge their lateral moraines unite to form a ***medial moraine***. Debris that has fallen down crevasses and become embedded in the ice is termed an ***englacial moraine***.

mountain natural upward projection of the Earth's surface, higher and steeper than a hill. The process of mountain building (orogeny) consists of volcanism, folding, faulting, and thrusting, resulting from the collision and welding together of two tectonic plates (see ◊plate tectonics). This process deforms the rock and compresses the sediment between the two plates into mountain chains.

mudflow downhill movement (mass movement) of muddy sediment containing a large proportion of water. Mudflows can be fast and destructive: in 1966 coal waste saturated with water engulfed a school in Aberfan, S Wales, killing 116 children. A ◊lahar is a form of mudflow associated with volcanic activity.

multinational corporation company or enterprise operating in several countries. Multinational corporations may be attracted to ◊developing countries because of cheap labour and low operating costs. They have been criticized on the grounds that they remove resources and profits with little concern for the development of the country concerned.

multiple-nuclei theory in the social sciences, a model of urban land use in which a city grows from several independent points rather than from one central business district. Each point acts as a ◊growth pole for a particular kind of land use, such as industry, retail, or high-quality housing. As these expand, they merge to form a single urban area.

The multiple-nuclei theory, developed by D Harris and E L Ullman in 1945, is the most complicated of the three standard ◊urban land-use models and the only one that gives some insight into the growth of cities in the developing world.

multiplier effect process whereby one change sets in motion a sequence of events that result in decline or growth. For example, in S Wales, pit closures in the coal industry resulted in unemployment, depopulation, closure of services, and disinvestment. This in turn led to further unemployment. Multiplier effects are also important in ◊new towns, where industry is needed to attract people and create further wealth.

N

national grid the network by which electricity is generated and distributed over a region or country. It contains many power stations and switching centres and allows, for example, high demand in one area to be met by surplus power generated in another. Britain has the world's largest grid system, with over 140 power stations able to supply up to 55,000 megawatts.

national park land set aside and conserved for public enjoyment. The first was Yellowstone National Park, USA, established 1872. National parks include not only the most scenic places, but also places distinguished for their historic, prehistoric, or scientific interest, or for their superior recreational assets. They range from areas the size of small countries to pockets of just a few hectares.

In England and Wales under the National Park Act 1949 ten national parks were established including the Peak District, the Lake District, and Snowdonia. National parks are protected from large-scale development, but from time to time pressure to develop land for agriculture, quarrying, or tourism, or to improve amenities for the local community means that conflicts of interest arise between landusers.

Other protected areas include Areas of Outstanding Natural Beauty and Sites of Special Scientific Interest (SSSIs).

natural arch see ◊arch.

natural gas mixture of flammable gases found in the Earth's crust, often in association with oil. It is one of the world's three main fossil fuels (with coal and oil). Natural gas is usually transported from its source by pipeline, although it may be liquefied for transport and storage and is, therefore, often used in remote areas where other fuels are scarce and expensive.

In the UK from the 1970s natural gas from the North Sea has superseded coal gas, or town gas, both as a domestic fuel and as an energy source for power stations.

natural increase the rise in population caused by ◊birth rate exceeding ◊death rate. Rates of natural increase vary considerably throughout the world. The highest rates are found in poor countries, but with industrialization they undergo ◊demographic transition. Natural increase excludes any population change due to migration.

nature reserve area set aside to protect a habitat and the wildlife that lives within it, with only restricted admission for the public. A nature reserve often provides a sanctuary for rare species. The world's largest is Etosha Reserve, Namibia; area 99,520 sq km.

In Britain, there are both officially designated nature reserves (such as Sites of Special Scientific Interest, managed by English Nature and the Scottish and Welsh Countryside Commissions, and those run by a variety of voluntary conservation organizations.

neighbourhood unit in the UK, an area within a ◊new town planned to serve the local needs of families. These units typically have a primary school, a ◊low-order shopping centre, a church, and a pub. Main roads form the boundaries. Neighbourhood units were designed to give a sense of community to people migrating to the new towns.

network system of ◊nodes (junctions) and links (transport routes) through which goods, services, people, money, or information flow. Networks are often shown on ◊topological maps.

neve another name for ◊firn.

newly industrialized country (NIC) country that has in recent decades experienced a breakthrough into manufacturing and rapid export-led economic growth. The prime examples are Taiwan, Hong Kong, Singapore, and South Korea. Their economic development during the 1970s and 1980s was partly due to a rapid increase of manufactured goods in their exports.

new town centrally planned urban area. In the UK, new towns such as Milton Keynes and Stevenage were built after World War II largely to accommodate the overspill from cities and large towns at a time when the population was rapidly expanding and inner-city centres had either decayed or been destroyed.

New towns are characterised by a regular street pattern and the presence of a number of self-contained neighbourhood units, consisting of houses, shops, and other local services. Modern industrial estates are located on the outskirts of towns where they are well served by main roads and motorways. Some new towns have been criticized as lacking character and for their failure to provide a stable industrial base. Since 1976 no new towns have been built in the UK.

nitrate pollution the contamination of water by nitrates. Nitrates in the soil, whether naturally occurring or from agricultural fertilizers, are used by plants to make proteins. However, increased use of artifical fertilizers means that higher levels of nitrates are being washed from the soil into rivers, lakes, and aquifers. There they cause an excessive enrichment of the water (◊eutrophication), leading to a rapid growth of algae, which in turn darkens the water and reduces its oxygen content. The water is expensive to purify and many plants and animals die.

nivation complex of physical processes, operating beneath or adjacent to snow, believed to be responsible for the enlargement of the hollows in which snow collects. It is also thought to play a role in the early formation of ◊corries. The processes involved include freeze-thaw (weathering by the alternate freezing and melting of ice), mass movement (the downhill movement of substances under gravity), and erosion by meltwater.

node point where routes meet. It may therefore be the same as a ◊route centre. In a topological ◊network, a node may be the start or crossing point of routes, also called a ***vertex***.

nomadic pastoralism ◊farming system where animals (cattle, goats, camels) are taken to different locations in order to find fresh pastures. It is practised in the developing world; for example, in central Asia and the Sahel region of W Africa. Increasing numbers of cattle may lead to overgrazing of the area and ◊desertification.

The increasing enclosure of land has reduced the area available for nomadic pastoralism and, as a result, this system of farming is under threat. The movement of farmers in this way contrasts with ◊sedentary agriculture.

nonrenewable resource natural resource, such as coal or oil, that takes thousands or millions of years to form naturally and can therefore not be replaced once it is consumed. The main energy sources used by humans are nonrenewable; ◊renewable sources, such as solar, tidal, and geothermal power, have so far been less exploited.

North Atlantic Drift warm ◊ocean current in the N Atlantic Ocean; an extension of the ◊Gulf Stream. It flows east across the Atlantic and has a mellowing effect on the climate of NW Europe, particularly the British Isles and Scandinavia.

nuclear energy or ***atomic energy*** energy released from the inner core, or nucleus, of the atom. Energy produced by nuclear fission (the splitting of uranium or plutonium nuclei) has been harnessed since the 1950s to generate electricity, and research continues into the possible controlled use of nuclear fusion (the fusing, or combining, of atomic nuclei).

In nuclear power stations, fission takes place in a ***nuclear reactor***. The nuclei of uranium or, more rarely, plutonium are induced to split, releasing large amounts of heat energy. The heat is then removed from the core of the reactor by circulating gas or water, and used to produce the steam that drives alternators and turbines to generate electrical power.

Unlike fossil fuels, such as coal and oil, which must be burned in large quantities to produce energy, nuclear fuels are used in very small amounts and supplies are therefore unlikely to be exhausted in the foreseeable future. However, the use of nuclear energy has given rise to concern over safety. Anxiety has been heightened by accidents such as the one at Chernobyl, Ukraine, in 1986, in which an explosive leak from a nuclear reactor caused clouds of radioactive material to be spread as far as Sweden. 31 people were killed in the explosion (many more are expected to die or become ill because of the long-term effects of radiation), and thousands of square kilometres of land were contaminated by fallout.

There has also been mounting concern about the production and disposal of toxic nuclear waste, which may have an active life of several thousand years, and the cost of maintaining nuclear power stations and decommissioning them at the end of their lives. In 1989, the UK government decided to postpone the construction of new nuclear power stations.

nucleated settlement settlement where the buildings are grouped or clustered around a central point, or nucleus. For example, a water supply such as a spring might form a central point for the initial development of a settlement; later development is often more linear.

Other types of settlement are ◊dispersed settlements and ◊linear developments.

nuée ardente glowing white-hot cloud of ash and gas emitted by a volcano during a violent eruption. In 1902 a nuée ardente produced by the eruption of Mount Pelee in Martinique swept down the volcano in a matter of seconds and killed 28,000 people in the nearby town of St Pierre.

nunatak mountain peak protruding through an ice sheet. Such peaks are common in Antarctica.

O

oasis area of land made fertile by the presence of water near the surface in an otherwise arid region. The occurrence of oases affects the distribution of plants, animals, and people in the desert regions of the world.

oat cereal plant grown in cool temperate regions. Oats are relatively little used as human food as they cannot be used for making bread and their grains have a high proportion of husk that is not easily removed. However, rolled oats and oatmeal are eaten as porridge and breakfast cereals.

In Europe, the importance of oats as an animal feed has diminished because of the rapid decline of the working horse population, and greater preference for higher-yielding barley.

occluded front weather ◊front formed when a cold front catches up with a warm front. It brings cloud and rain as air is forced to rise upwards along the front, cooling and condensing as it does so.

ocean great mass of salt water. Strictly speaking three oceans exist – the Atlantic, Indian, and Pacific – to which the Arctic is often added. Oceans cover approximately 70% of the total surface area of the Earth.

ocean current fast-flowing current of seawater generated by the wind or by variations in water density between two areas. Ocean currents are partly responsible for transferring heat from the equator to the poles and thereby evening out the global heat imbalance.

The ◊Gulf Stream is an example of a warm ocean current; it flows from the Gulf of Mexico to NW Europe and is responsible for keeping the coast of Scandinavia free of ice in winter. The Peru Current is a cold ocean current, transferring cold water from the Antarctic to the coasts of Chile and Peru. Small fish called anchovies thrive in these fertile waters, providing an important source of income for Peru. At approximate five-to-eight-year intervals, the phenomenon of ◊El Niño causes the Peru Current to become warm and has disastrous results (as in 1982–83) for Peruvian wildlife and for the anchovy industry.

ocean ridge mountain range on the seabed indicating the presence of a ◊constructive margin (where tectonic plates are moving apart and magma rises to the surface). An ocean ridge can rise thousands of metres above the surrounding seabed.

Ocean ridges, such as the ◊Mid-Atlantic Ridge, consist of many segments offset along faults (see ◊faulting).

ocean trench deep trench in the seabed indicating the presence of a destructive margin (produced by the movements of ◊plate tectonics). Ocean trenches are found around the edge of the Pacific Ocean. They represent the deepest parts of the ocean floor, the deepest being the ◊Mariana Trench, off Japan, which has a depth of 11,034 m.

oil crop plant from which vegetable oils are pressed from the seeds. Cool temperate areas, such as the UK, grow rapeseed and linseed; warm temperate regions produce sunflowers, olives, and soya beans; tropical regions produce groundnuts (peanuts), palm oil, and coconuts. Some of the major vegetable oils, such as soya bean oil, peanut oil, and cottonseed oil, are derived from crops grown primarily for other purposes. Most vegetable oils are used as both edible oils and as ingredients in industrial products such as soaps, varnishes, printing inks, and paints.

oil, crude thick flammable mineral oil found underground in permeable rocks; see ◊petroleum.

oil refinery industrial complex where crude oil is processed into different products. The light volatile parts of the oil form ◊petroleum, while the heavier parts make bitumen and petrochemicals. Oil refineries are often located at deep-water ports or near their industrial markets. They need flat land and water for cooling.

oilseed rape plant grown for its seeds, which yield an edible oil used in the production of margarine. Oilseed rape is a member of the mustard family and has bright yellow flowers.

OPEC acronym for ***Organization of Petroleum-Exporting Countries*** a body established 1960 to protect the interests and avoid the exploitation of certain oil-producing countries, notably Iran, Iraq, Kuwait, Saudi Arabia, Venezuela, the United Arab Emerates, Libya, and Nigeria. It meets regularly to set oil production levels and coordinate prices.

Not all oil-producing countries are members of OPEC; for example, the UK is not a member.

opencast mining mining at the surface rather than underground. Coal, iron ore, and phosphates are often extracted by opencast mining. Often the mineral deposit is covered by soil, which must first be stripped off, usually by large machines such as walking draglines and bucket-wheel excavators. The ore deposit is then broken up by explosives.

Ordnance Survey (OS) official body, established 1791, responsible for the mapping of Britain. Maps are produced to a variety of scales, the most commonly used being 1:50,000 and 1:25,000.

In 1989, the OS began using a computerized system for the creation and continuous revision of maps. Customers can now have maps drafted to their own specifications, choosing from over 50 features (such as houses, roads, and vegetation).

ore body of rock or a deposit of sediment that is worth mining for the economically valuable mineral it contains. The term is usually applied to sources of metals. For example, iron ores contain minerals such as iron oxides and carbonates that can be reduced to iron, usually by smelting in a ◊blast furnace.

organic farming farming without the use of artificial fertilizers, pesticides (such as weedkillers or insecticides), or other agrochemicals (such as hormones, growth stimulants, or fruit regulators).

In place of artificial fertilizers, compost, manure, seaweed, or other substances derived from living things are used (hence the name 'organic'). Growing a crop of a nitrogen-fixing plant (◊legume), then ploughing it back into the soil, also fertilizes the ground. Weeds can be controlled by hoeing, mulching (covering with manure, straw, or black plastic), or burning off. Organic farming methods produce food without pesticide residues and greatly reduce pollution of the environment. They are more labour intensive and therefore more expensive, but use less fossil fuel.

orogeny or ***orogenesis*** the formation of mountains. It is brought about by the movements of the rigid plates making up the Earth's crust (described by ◊plate tectonics). Where two plates collide at a destructive margin rocks become folded and lifted to form chains of fold mountains (such as the ◊young fold mountains of the Himalayas).

orographic rainfall rainfall that occurs when an airstream is forced to rise over a mountain range. As the air rises, it becomes cooled. The amount of moisture that air can hold decreases with decreasing temperature. So the water vapour in the rising airstream condenses, and rain falls on the windward side of the mountain. The air descending on the leeward side contains less moisture, resulting in a ***rainshadow***, where there is little or no rain.

In the UK the Pennine hills, which extend southwards from Northumbria to Derbyshire in N England, interrupt the path of the prevailing southwesterly winds, causing orographic rainfall. Their presence is partly responsible for the west of the UK being wetter than the east. The orographic effect can sometimes occur in large cities, when air rises over tall buildings.

orographic rainfall

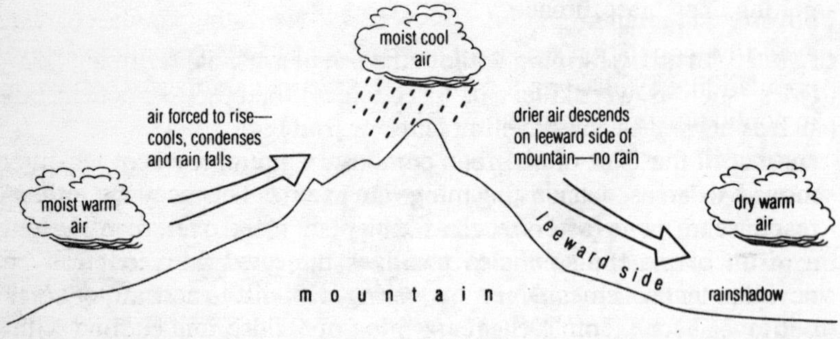

outwash sands and gravels deposited by streams of meltwater (water produced by the melting of a glacier). Such material may be laid down ahead of the glacier's snout to form a large flat expanse called an ***outwash plain***.

Outwash is usually well sorted, the particles being deposited by the meltwater according to their size – the largest are deposited next to the snout while finer particles are deposited further downstream.

overfishing ◊fishing at rates that exceed the natural replacement rates of fish species, resulting in a net population decline. For example, in the North Atlantic, herring has been fished to the verge of extinction and the cod and haddock populations are severely depleted. In the developing world, use of huge factory ships, often by Western fisheries, has depleted stocks for local people who cannot obtain protein in any other way.

overland flow another term for ◊surface runoff of water after rain.

overpopulation too many people for the resources available in an area (such as food, land, and water). The consequences were first set out in the ◊Malthus theory.

Although there is often a link between overpopulation and ◊population density, high densities will not always result in overpopulation. In many countries, resources are plentiful and the ◊infrastructure and technology are well developed. This means that a large number of people can be supported by a small area of land. In some developing countries, such as Bangladesh, Ethiopia, and Brazil, insufficient food, minerals, and energy, and inequitable income distribution result in poverty and often migration in search of better living conditions. Here even low population densities may amount to overpopulation. Overpopulation may also result from a decrease in resources or an increase in population or a combination of both.

overspill that part of an urban population or employment which has moved to areas outside the main settlement. This may be either in response to ◊push factors, such as ◊congestion and overcrowding, or as a result of planning policies perhaps connected with inner-city redevelopment schemes. ◊New towns may be built to accommodate this outflow, as for example, Stevenage, Hertfordshire, and Harlow, Essex.

oxbow lake

the formation of an oxbow lake

A

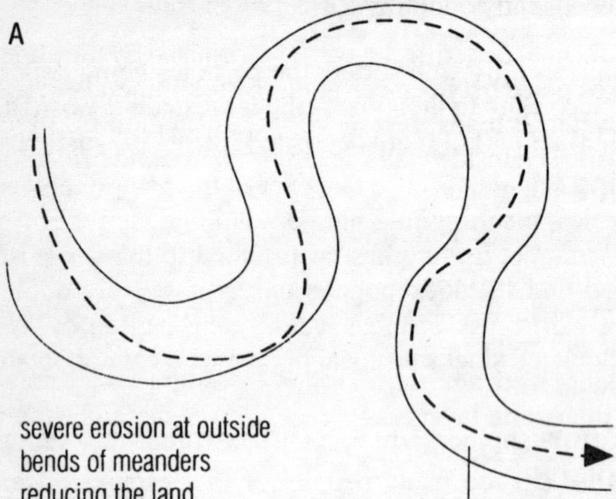

B

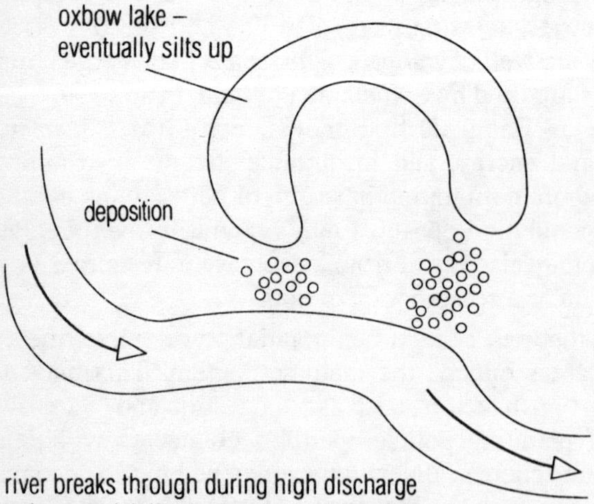

oxbow lake curved lake found on the flood plain of a river. Oxbows are caused by the loops of ◊meanders being cut off at times of flood and the river subsequently adopting a shorter course.

oxidation form of ◊chemical weathering caused by the chemical reaction that takes place between certain iron-rich minerals in rock and the oxygen in water. It tends to result in the formation of a red-coloured soil or deposit. The inside walls of canal tunnels and bridges often have deposits formed in this way.

ozone O_3 highly reactive pale-blue gas; a type of oxygen. It forms a layer in the upper atmosphere, which protects life on Earth from ultraviolet rays, a cause of skin cancer. At lower atmospheric levels it is an air pollutant and contributes to the ◊greenhouse effect.

A continent-sized hole has formed over Antarctica as a result of damage to the ozone layer, caused in part by the emission of chlorofluorocarbons (CFCs).

P

palaeomagnetism the reconstruction of the Earth's ancient magnetic field. Palaeomagnetism has been used to demonstrate ◊continental drift by determining the direction of the magnetic field of dated rocks from different continents.

Pampas in South America, a flat treeless plain lying between the Andes mountain range and the Atlantic and rising gradually from the coast to the lower slopes of the mountains. The Pampas contains large cattle ranches although some areas are arid and unproductive.

Pangaea ancient single land mass, made up of all the present continents, believed to have existed between 250 and 200 million years ago; the rest of the Earth was covered by the Panthalassa ocean. Pangaea split into two land masses – ◊Laurasia in the north and ◊Gondwanaland in the south – which subsequently broke up into several continents. These then drifted slowly to their present positions (see ◊continental drift).

park and ride town-planning scheme in which parking space is provided (often free) some distance away from the central business district. Shoppers are taken by bus to the central area, which may be traffic-free (◊pedestrianization). Park and ride is one of the planning strategies that can be used to combat ◊congestion. Park and ride became widespread in the 1980s; in the UK, an example is in Oxford.

pastoral farming the rearing or keeping of animals in order to obtain meat or other products, such as milk, skins, and hair. Animals can be kept in one place or periodically moved (◊nomadic pastoralism).

peasant country-dweller engaged in small-scale farming. A peasant normally owns or rents a small amount of land, working with an aim to be self-sufficient and to sell surplus supplies locally.

In the UK, the move towards larger farms in the 18th century resulted in the disappearance of the independent peasantry, although small-scale farming survives in smallholdings and Scottish crofts. However, in many countries the tradition of small independent landholding remains a distinctive way of life today.

peat fibrous organic substance found in bogs and formed by the incomplete decomposition of plants such as sphagnum moss. N Asia, Canada, Finland, Ireland, and other places have large deposits, which have been dried and used as fuel from ancient times. Peat can also be used as a soil additive.

Peat bogs began to be formed when glaciers retreated, about 9,000 years ago. They grow at the rate of only a millimetre a year, and large-scale digging can result in destruction both of the bog and of specialized plants growing there. Some 6% of land area in Britain is covered by peat bog, extending to depth of about 3 metres.

pedestrianization the closing of an area to traffic, making it more suitable for people on foot. It is now common in many town shopping centres, since cars and people often obstruct one another. This restricts ◊accessibility and causes ◊congestion. Sometimes service vehicles (such as buses and taxis) are allowed access.

peninsula land surrounded on three sides by water but still attached to a larger landmass. Florida, USA, is an example.

periglacial bordering a glacial area but not actually covered by ice. Periglacial areas today include parts of Siberia, Greenland, and North America. The soil in these areas is frozen to a depth of several metres (◊permafrost) with only the top few centimetres thawing during the brief summer. The vegetation is characteristic of ◊tundra.

During the last ice age all of southern England was periglacial. Weathering by ◊freeze-thaw (the alternate freezing and thawing of ice in rock cracks) would have been severe, and ◊solifluction would have taken place on a large scale, causing wet topsoil to slip from valley sides.

periphery outlying area with backward economy (see ◊core and periphery).

permafrost condition in which a deep layer of soil does not thaw out during the summer. Permafrost occurs under ◊periglacial conditions. It is claimed that 26% of the world's land surface is permafrost.

Permafrost gives rise to a poorly drained form of grassland typical of N Canada, Siberia, and Alaska known as ◊tundra.

permanent pasture area of grassland that has consistently been given over to feeding cattle or livestock. These areas ideally have deep soil, access to water (near a river, for example), and a long growing season.

permeable rock rock that allows water to pass through it. Rocks are permeable if they have cracks or joints running through them or if they are porous, containing many interconnected pores. Examples of permeable rocks include limestone (which is heavily jointed) and chalk (porous).

Unlike ◊impermeable rocks, which do not allow water to pass through, permeable rocks rarely support rivers and are therefore subject to little erosion. As a result they commonly form upland areas (such as the chalk downs of SE England, and the limestone Pennines of N England).

pervious rock another name for ◊permeable rock.

petrochemical chemical derived from the processing of petroleum (crude oil). ***Petrochemical industries*** are those that obtain their raw materials from the processing of petroleum.

petroleum or ***crude oil*** natural mineral oil, a thick greenish-brown flammable liquid formed underground by the decomposition of organic matter. Oil may flow naturally from wells under gas pressure from above or water pressure from below, or it may require pumping to bring it to the surface – for example, new technologies have been introduced to pump oil from offshore wells (for example, in the North Sea) and from the Arctic (the Alaska pipeline).

A number of products can be made from petroleum – for example, fuel oil, petrol (gasoline), diesel, paraffin wax, and petroleum jelly. These and other petrochemicals are used in large quantities for the manufacture of detergents, artificial fibres, plastics, pesticides, fertilizers, and drugs.

The burning of fuels derived from petroleum (notably by cars) is a major cause of air pollution, and oil spills from tankers and drilling rigs have been the cause of a number of environmental catastrophes.

pH scale for measuring acidity or alkalinity. A pH of 7.0 indicates neutrality, below 7 is acid, while above 7 is alkaline. ◊Acid rain, caused by airborne pollutants such as sulphur dioxide and nitrous oxides, has a pH of less than 5.0; the average pH of rain in eastern England is 4.3.

physical weathering or ***mechanical weathering*** form of ◊weathering responsible for the mechanical breakdown of rocks but involving no chemical change. Processes involved include ◊freeze-thaw (the alternate freezing and melting of ice in rock cracks) and ◊exfoliation (the alternate expansion and contraction of rocks in response to extreme changes in temperature).

pie chart method of displaying proportional information by dividing a circle up into different-sized sectors (slices of pie). The angle of each sector is proportional to the size, expressed as a percentage, of the item of data that it represents.

To construct a pie chart:

(1) convert each item of data to a percentage figure;

(2) 100% will equal the whole circle (360°), therefore each 1% will equal 360°/100, or 3.6°;

pie chart

traffic survey of the types of vehicle passing a certain point in five minutes

vehicle type	*number in 5 min*	*percentage*	*angle of pie-chart sector*
car	18	18/28 x 100 = 64.3%	64.3% x 3.6 = 231°
bus	2	2/28 x 100 = 7.1%	7.1% x 3.6 = 26°
lorry	1	1/28 x 100 = 3.6%	3.6% x 3.6 = 13°
van	4	4/28 x 100 = 14.3%	14.3% x 3.6 = 51°
bicycle	3	3/28 x 100 = 10.7%	10.7% x 3.6 = 39°
total	28	100.0%	360°

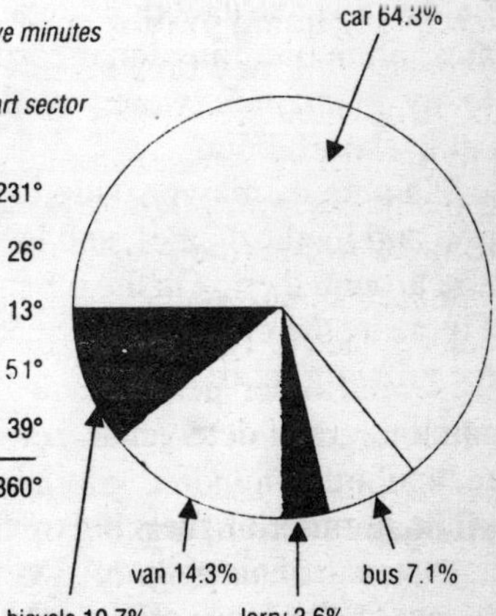

(3) calculate the angle of the segment for each item of data, and plot this on the circle.

The diagram may be made clearer by adding colours or shadings to each group, together with a key.

pipeflow the movement of water through pipes in the soil; it is a form of ◊throughflow and may be very rapid. The pipes used may be of animal or plant origin – for example, worm burrows or gaps created by tree roots.

placer deposit concentration of an economically important mineral, such as gold, in sediments laid down by wind or water. Examples are the Witwatersrand gold deposits of South Africa, which are gold- and uranium-bearing sediments laid down by ancient rivers.

plain large area of flat land, usually covered with grass. Plains cover a large proportion of the Earth's surface, especially between the deserts of the tropics and the rainforests of the equator, and have rain in one season only. Examples are the North European Plain, the High Plains of the USA and Canada, and the Russian Plain (also known as the steppe).

plantation large farm or estate where commercial production of one crop – such as rubber (in Malaysia), palm oil (in Nigeria), or tea (in Sri Lanka) – is carried out. Plantations are usually owned by large companies, often ◊multinational corporations, and run by an estate manager. Many plantations were established in countries under colonial rule, using slave labour.

Plantations may provide local jobs, extra facilities (such as education and medical care), and lead to the development of the ◊infrastructure around them. On the other hand, they may disadvantage an area by exporting the profits and by taking up land that could otherwise be used for growing food.

plate according to plate tectonics, one of a number of slabs of solid rock, about a hundred kilometres thick and often several thousands of kilometres across, making up the Earth's surface.

Plates are made up of two types of crustal material: oceanic crust (sima) and continental crust (sial). ***Oceanic crust*** is heavy and consists largely of basalt. It is formed at constructive margins. ***Continental crust*** is less dense and is rich in granite. Made up of volcanic islands and folded sediments, it is usually associated with destructive margins.

plateau elevated area of fairly flat land, or a mountainous region in which the peaks are at the same height. Examples are the Tibetan Plateau and the Massif Central in France.

plate tectonics theory that explains the formation of the major physical features of the Earth's surface. The Earth's outermost layer, or lithosphere, is regarded as being divided into a number of rigid plates up to a hundred kilometres thick, which move relative to each other. Their movement may be due to convection currents within the semi-solid mantle beneath. Major landforms occur at plate margins (see ◊constructive margin and ◊destructive margin) – for example, volcanoes, young fold mountains, ocean trenches, and ocean ridges.

plate tectonics

the plates of the Earth's lithosphere

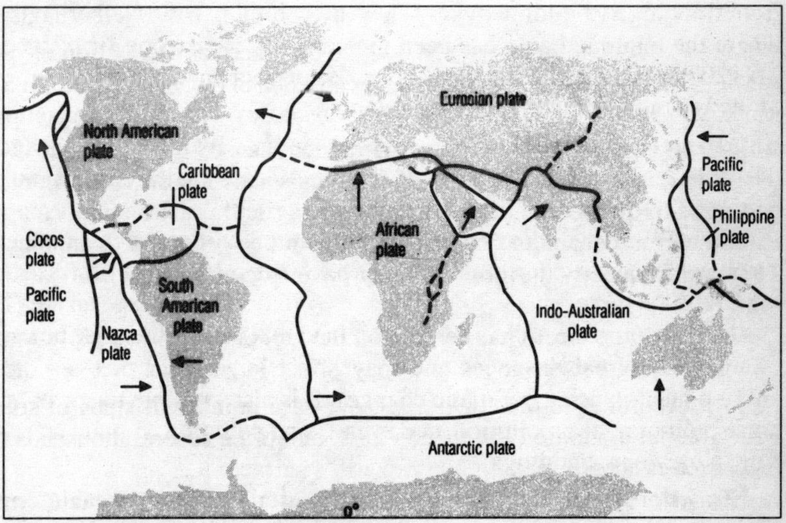

—— plate margins - - - - uncertain plate margins

Pleistocene in ◊geological time, the first epoch of the Quarternary period. It began 1.8 million years ago and ended 10,000 years ago. Glaciers were abundant during the ice age of this period.

plough agricultural implement used for tilling the soil. The plough dates from about 3500 BC, when oxen were used to pull a simple wooden blade.

plucking a process of glacial erosion. Water beneath a glacier will freeze fragments of loose rock to the base of the ice. When the ice moves, the rock fragment is 'plucked' away from the underlying bedrock. Plucking is thought to be responsible for the formation of steep, jagged slopes such as the backwall of the corrie and the downslope-side of the roche moutonnée.

plunge pool deep pool at the bottom of a ◊waterfall. It is formed by the hydraulic action of the water as it crashes down onto the river bed from a height.

polder area of flat reclaimed land that used to be covered by a river, lake, or the sea. Polders have been artificially drained and protected from flooding by building dykes. They are common in the Netherlands, where the total land area has been increased by nearly one-fifth since AD 1200. Such schemes as the Zuider Zee project have provided some of the best agricultural land in the country.

pollution contamination of the environment caused by human activities. It frequently takes the form of chemicals added to the land, water, or air as a by-product of industry, traffic, or agriculture. Pollutants may enter the ◊food chain and be passed on from one organism to another. They are frequently harmful and may have side effects such as ◊acid rain.

Much recent concern has centred on the fact that chemical pollution often travels great distances and may affect large areas or even the entire planet (causing ◊climatic change). It is also possible to speak of noise pollution, heat pollution, and even visual pollution (usually referring to ugly new buildings).

pools and riffles alternating deeps (pools) and shallows (riffles) along the course of a river. There is evidence to suggest a link between pools and riffles and the occurrence of ◊meanders (bends in a river), although it is not certain whether they are responsible for meander formation.

population the number of people inhabiting a country, region, area, or town. Population statistics are derived from many sources, for example through the registration of births and deaths; and from censuses of the population.

population control measures taken by some governments to limit the growth of their countries' populations by trying to reduce ◊birth rates. Propaganda, freely available contraception, and tax disincentives for large families are some of the measures that have been tried.

The population-control policies introduced by the Chinese government are the best known. In 1979 the government introduced a 'one-child policy' that encouraged ◊family planning and penalized couples who have more than one child. It has been only partially successful since it has been difficult to administer, especially in rural areas, and has in some cases led to the killing of girls in favour of sons as heirs.

population density the number of people living in a given area, usually expressed as people per square kilometre. It is calculated by dividing the population of a region by its area.

Population density provides a useful means for comparing ◊population distribution. Densities vary considerably over the globe. High population densities may amount to ◊overpopulation only where resources are scarce.

population distribution the location of people within an area. Areas may be compared by looking at variations in ◊population density. Population is unevenly distributed for a number of reasons. ◊Pull factors (which attract people) include mineral resources, temperate climate, the availability of water, and fertile, flat land. ◊Push factors (which repel people) include dense vegetation, limited accessibility, and political or religious oppression.

Over 90% of the world's population is found in the northern hemisphere, where there is most land. Some 60% of the world's land surface is unpopulated. At the other extreme, urban areas such as Mong Kok in

Hong Kong may have as many as 160,000 people living in one square kilometre.

population explosion the rapid and dramatic rise in world population that has occurred over the last few hundred years. Between 1959 and 1990, the world's population increased from 2.5 billion to over 5 billion people. It is estimated that it will be at least 6 billion by the end of the century. Most of this growth is now taking place in the ◊developing world, where rates of natural increase are much higher than in rich countries. Concern that this might lead to ◊overpopulation has led some countries to adopt policies of ◊population control.

population pyramid graph of the population of an area, such as a country or town, using age and sex groupings. The pyramid is arranged with the youngest age group at the bottom and the eldest at the top, at intervals of five to ten years. The pyramid is split down the middle with one side showing male groups and the other side female. Population pyramids are a useful way of summarizing the demographic characteristics of an area. Their shape reflects changes in ◊birth rate and ◊death rate.

porous rock rock containing a large number of small holes, or pores; for example chalk and sandstone. Many porous rocks are also permeable, allowing water to pass through.

port point where goods are loaded or unloaded from a water-based to a land-based form of transport. Most ports are coastal, though inland ports on rivers also exist. Ports often have specialized equipment to handle cargo in large quantities (for example, ◊container or ◊roll on/roll off facilities).

Historically, ports have been important ◊growth poles from which the transport networks of many colonial and trading countries developed, as in Nigeria. Ports with deep-water berths can accommodate large modern shipping; for example, the port of Rotterdam. See also ◊airport.

pothole small hollow in the rock bed of a river. Potholes are formed by the erosive action of rocky material carried by the river (corrasion), and are commonly found along the river's upper course, where it tends to flow directly over solid bedrock.

population pyramid

population pyramid for India (1981)

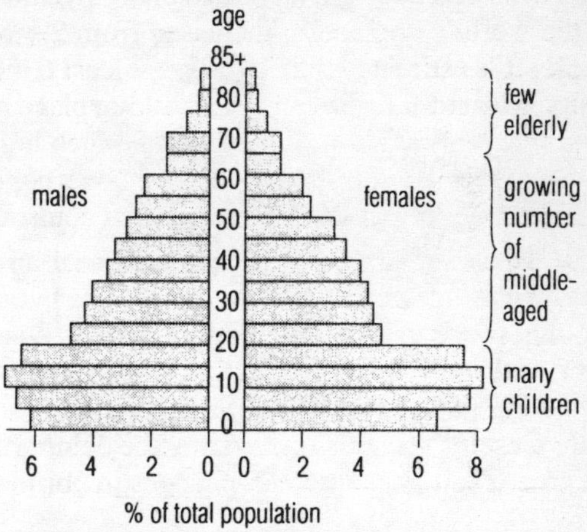

population pyramid for the UK (1981)

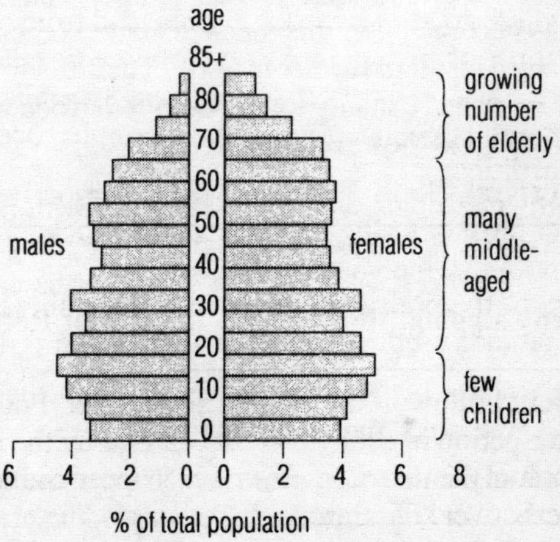

poverty cycle

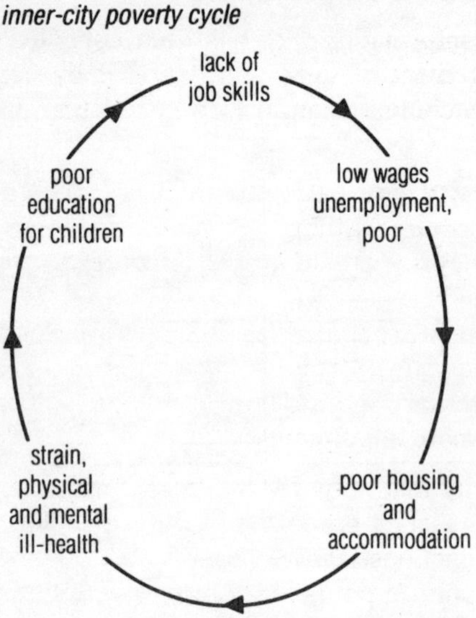

poverty cycle set of factors or events by which poverty, once started, is likely to continue unless there is outside intervention. Once an area or a person has become poor, this tends to lead to other disadvantages, which may in turn result in further poverty. The situation is often found in ◊inner-city areas and ◊shanty towns. Applied to countries, the poverty cycle is often called the ***development trap***.

prairie the central North American plain, formerly grass-covered, extending over most of the region between the Rockies on the west and the Great Lakes and Ohio River on the east.

Large areas of the prairie are devoted to the cultivation of cereals; the farming methods used are typical of ◊extensive agriculture.

precipitation water that falls to the Earth from the atmosphere. It includes rain, snow, sleet, hail, dew, and frost.

prevailing wind the direction from which the wind most commonly blows in a locality. In NW Europe, for example, the prevailing wind is southwesterly, blowing from the Atlantic Ocean in the southwest and bringing moist and warm conditions.

primary data information that has been collected at first hand. It involves measurement of some sort, whether by taking readings off instruments, sketching, counting, or conducting interviews (using questionnaires).

primary industry any extractive industry, including mining and quarrying. Agriculture, fishing, and forestry are also included in this category since they involve the extraction of resources (see ◊industrial sectors).

Developing countries often have a higher proportion of their workforce involved in primary industries than developed countries, where secondary and tertiary industries (see ◊secondary industry, ◊tertiary industry) are of more importance.

primate city city that is by far the largest within a country or area. Such a city holds a larger proportion of the population, economic activity, and social functions than other settlements within that area. It is also likely to dominate politically together with the surrounding core area (see ◊core and periphery).

Not all urban hierarchies are arranged in this way, but it is common in developing countries, where ◊migration to the primate city may be very rapid (sometimes forming ◊millionaire cities). Primate cities are also found in developed countries; London and Paris are examples.

pull factor any factor that tends to attract people to an area (see ◊migration). Examples include higher wages, better housing, and better educational opportunities.

pulse the edible seed of a leguminous plant (see ◊legume) – for example, peas and beans. Pulses provide a concentrated source of vegetable protein, and make a vital contribution to human diets in poor countries where meat is scarce. They also play a useful role in ◊crop rotation as they help to raise soil nitrogen levels.

Soya beans are the major temperate protein crop in the West; most are used for oil production or for animal feed. In Asia, most are processed into soya milk and beancurd. Groundnuts (peanuts) dominate pulse production in the tropical world and are generally consumed as human food.

pumice light extrusive, or volcanic, rock produced by the frothing action of expanding gases during the solidification of lava. It has the texture of a hard sponge and is used as an abrasive.

pumped storage hydroelectric plant that uses surplus electricity to pump water back into a high-level reservoir. In normal working the water flows from this reservoir through the turbines to generate power for feeding into the grid. At times of low power demand, electricity is taken from the grid to turn the turbines into pumps that then pump the water back again. This ensures that there is always a maximum 'head' of water in the reservoir to give the maximum output when required.

An example of a pumped-storage plant is in Dinorwig, North Wales.

push factor any factor that tends to repel people from an area (see ◊migration). Examples are crop failure, pollution, natural disasters, and poor living standards.

pyramidal peak angular mountain peak found in glaciated areas; for example, the Matterhorn in Switzerland. It is formed when three or four ◊corries (steep-sided hollows) are eroded, back-to-back, around the sides of a mountain, leaving an isolated peak in the middle.

pyroclastic deposit deposit made up of fragments of rock, ranging in size from fine ash to large boulders, ejected during an explosive volcanic eruption.

Q

quartz crystalline mineral found in many different kinds of rock, including sandstone and granite. It is one of the most abundant minerals of the Earth's crust (12% by volume). Quartz is very hard, and is resistant to chemical and physical weathering.

quartzite ◊metamorphic rock consisting of pure quartz sandstone that has recrystallized under increasing heat and pressure.

Quaternary period of geological time that began about 2 million years ago and is still in process. It is divided into the Pleistocene and Holocene epochs.

quaternary industry any employment that may be regarded as professional; for example, accountancy, systems analysis, and research and development. Quaternary industries normally involve working in offices, and are most important in developed countries.

quota in international trade, a limitation on the amount of a commodity that may be exported, imported, or produced. Restrictions may be imposed forcibly or voluntarily.

The European Community's ◊Common Agricultural Policy has introduced quotas to limit the production of milk because rising herd yields and a trend towards a healthier low-fat diet have led to overproduction. Dairy farmers will not be paid for any milk produced above a set limit.

R

radiation heat energy given off by a warm body such as the Sun.

Only a tiny fraction of the Sun's radiation reaches the Earth: as it passes through the ◊atmosphere much of it is absorbed and scattered by dust, water vapour, and gases such as ozone. The amount that eventually reaches the surface, called ◊insolation, is greatest at the equator, where the Sun is most directly overhead and the radiation most concentrated. At the poles, where the angle of the Sun's rays is much lower, the insolation is less.

Radiation is not only received by the Earth, it is also given off by its surface (***ground radiation***). The climatic difference between continents and oceans is partly due to the fact that land loses and gains heat more rapidly than the sea.

radiation fog see ◊fog.

rain separate drops of water that fall to the Earth's surface from clouds. The drops are formed by the accumulation of droplets that condense from water vapour around dust particles. Rain can form in three main ways (◊frontal rainfall, ◊orographic rainfall, and ◊convectional rainfall.)

rainfall gauge instrument used to measure ◊precipitation, usually rain. It consists of an open-topped cylinder, inside which there is a close-fitting funnel that directs the rain to a collecting bottle. The gauge may be partially embedded in soil to prevent spillage. The amount of water that collects in the bottle is measured every day, usually in millimetres. When the amount of water collected is too little to be measured ***trace rainfall*** is said to have taken place.

Snow falling into the gauge must be melted before a measurement is taken.

rainforest dense forest found on or near the ◊equator where the climate is hot and wet. Over half the tropical rainforests are in Central and

rainfall gauge

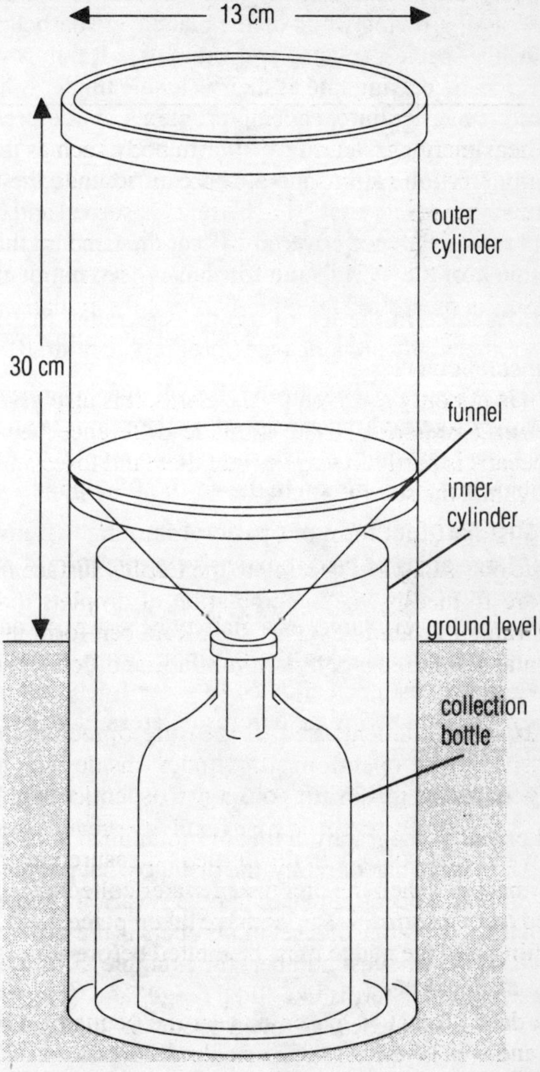

South America, the rest in SE Asia and Africa. Although covering approximately 6% of the Earth's land surface, rainforests comprise about 50% of all growing wood on the planet, and harbour at least 40% of the Earth's species (plants and animals). Rainforests are being destroyed at an increasing rate as their valuable timber is harvested and land cleared for agriculture, causing problems of ◊deforestation, ◊soil erosion, and flooding. Clearing of the rainforests may also lead to a global warming of the atmosphere, and contribute to the ◊greenhouse effect.

Tropical rainforest once covered 14% of the Earth's land surface. By 1991 over 50% of the world's rainforest had been removed .

rainshadow zone in the lee of a mountain range that receives less rainfall than the windward side (see ◊orographic rainfall).

raised beach beach that has been raised above the present-day shoreline and is therefore no longer washed by the sea. It is an indication of a fall in sea level (eustatic) or of a rise in land level (isostatic).

Raised beaches are common in the north of Scotland.

ranching commercial form of ◊pastoral farming that involves extensive use of large areas of land (◊extensive agriculture) for grazing cattle or sheep.

Ranches may be very large, especially where the soil quality is poor; for example, the estancias on the ◊Pampas grasslands in Argentina. Cattle have in the past been allowed to graze freely but more are now enclosed. In the Amazon some deforested areas have been given over to beef-cattle ranching.

rancho Venezuelan makeshift housing or ◊shanty town.

range in physical geography, a line of mountains (such as the Alps or Himalayas). In human geography, the distance that people are prepared to travel (often to a ◊central place) to obtain various goods or services. In mathematics, the range of a set of numbers is the difference between the largest and the smallest number; for example, 5, 8, 2, 9, 4 = 9 – 2 = 7; this sense is used in terms like 'tidal range' and 'temperature range'. Range is also a name for an open piece of land where cattle are ranched.

raw material any substance that forms the starting point of industrial or food processing; for example, coal, timber, and wheat. Raw materials are the result of primary ⇨industrial sectors, such as mining or agriculture, and form the inputs to secondary, manufacturing activities.

recreation any leisure activity that does not involve an overnight stay (this is tourism). Rambling, fishing, and picnicking are examples.

recycling processing industrial and household waste (such as paper, glass, and some metals and plastics) so that it can be reused. This saves expenditure on scarce raw materials, slows down the depletion of ⇨nonrenewable resources, and helps to reduce pollution.

Most British recycling schemes are voluntary, and rely on people taking waste items to a central collection point. However, some local authorities, such as Leeds, now ask householders to separate waste before collection, making recycling possible on a much larger scale.

refugee person fleeing from oppressive or dangerous conditions (such as political, religious, or military persecution) and seeking refuge in a foreign country. In 1991 there were an estimated 17 million refugees worldwide. A distinction is usually made by Western nations between 'political' refugees and so-called 'economic' refugees, who are said to be escaping from poverty rather than persecution, particularly when the refugees come from low-income countries.

regolith the surface layer of loose material that covers most rocks. It consists of eroded rocky material, volcanic ash, river alluvium, vegetable matter, or a mixture of these known as ⇨soil.

rejuvenation the renewal of a river's powers of downward erosion. It may be caused by a fall in sea level or a rise in land level, or by the increase in water flow that results when one river captures another (⇨river capture).

Several river features are formed by rejuvenation. For example, as a river cuts down into its channel it will leave its old floodplain perched up on the valley side to form a ⇨river terrace. Meanders (bends in the river) become deeper and their sides more steep, forming ⇨incised meanders. ⇨Waterfalls and rapids become more common (at the ⇨knickpoint).

rejuvenation

river features formed by rejuvenation

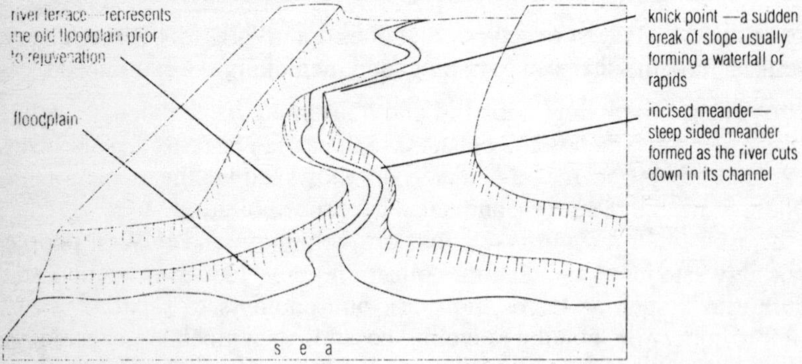

relief the ups and downs of a landscape. An area can be said to have undulating or flat relief. All areas have relief.

renewable energy power from any source that replenishes itself. Most renewable systems rely on ◊solar energy directly or through the weather cycle as ◊wave power, ◊hydroelectric power, or ◊wind energy, or solar energy collected by plants (alcohol fuels, for example). In addition, the gravitational force of the Moon can be harnessed through ◊tidal power stations, and the heat trapped in the centre of the Earth is used via ◊geothermal energy systems.

renewable resource natural resource that is replaced by natural processes in a reasonable amount of time. Properly conserved, soil, water, forests, plants, and animals are all renewable resources.

resettlement scheme policy for moving people (usually ◊squatters) out of their homes to other areas where they are provided with basic dwellings. This is common in urban areas of developing countries; for example, in Bombay, India. Resettlement on a large scale took place in Indonesia in the 1980s, and was known as ◊transmigration.

resort town urban settlement whose main function is to cater for tourism of one type or another. For example, coastal resorts (such as

Blackpool, Lancashire) offer seaside activities; Alpine resorts (such as Chamonix, France) offer skiing. Resorts may also have attractive climate or scenery, or historic interest. Often a high proportion of the workforce is engaged in tertiary activities associated with tourism (see ◊industrial sectors) and there may be seasonal unemployment.

resource commodity that can be used to satisfy human needs. Resources can be categorized into ***human resources***, such as labour, supplies, and skills, and ***natural resources***, such as climate, fossil fuels, and water. Natural resources may be further divided into ◊nonrenewable resources and ◊renewable resources.

The rich nations' demands for resources are causing concern among many people who feel that the present and future demands of industrial societies cannot be sustained for more than a century or two, and that this will be at the expense of the ◊developing world and the global environment. Other authorities believe that new technologies will emerge, enabling resources that are now of little importance to replace those being exhausted.

retail sale of goods and services to a consumer.

The large range of retail outlets include vending machines, street pedlars, specialized shops, department stores, supermarkets and hypermarkets, and cooperative stores. These are supplemented by auctions, door-to-door selling, telephone selling, and mail order.

ria long narrow sea inlet, usually surrounded by hills. It is formed by the flooding of a river valley due to either a rise in sea level or a lowering of a landmass. There are a number of rias in the UK – for example, in Salcombe and Dartmouth in Devon.

ribbon development another term for ◊linear development, housing that has grown up along a route.

ribbon lake long, narrow lake found on the floor of a ◊glacial trough. A ribbon lake will often form in an elongated hollow carved out by a glacier, perhaps where it came across a weaker band of rock. Ribbon lakes can also form when water ponds up behind a terminal moraine or a landslide. The English Lake District is named after its many ribbon lakes, such as Lake Windermere and Coniston Water.

ridge of high pressure

ridge of high pressure shown on a synoptic weather chart

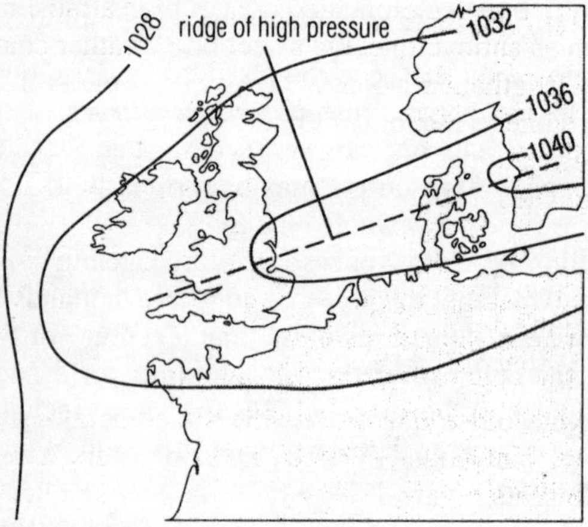

rice principal cereal of the wet regions of the tropics. It is unique among cereal crops in that it is grown standing in water. The yield is very large, and rice is said to be the staple food of one-third of the world population. New varieties with greatly increased protein content have been developed and yields are higher than ever before (see ◊green revolution).

Rice takes 150–200 days to mature in warm, wet conditions. During its growing period, it needs to be flooded either by the heavy ◊monsoon rains or by adequate ◊irrigation. Outside Asia, there is some rice production in the Po valley of Italy, and in the USA in Louisiana, the Carolinas, and California.

Richter scale scale used to measure the magnitude of an ◊earthquake (the ◊Mercalli scale, by contrast, measures the effects of an earthquake).

An earthquake's magnitude is a function of the total amount of energy, and each point on the Richter scale represents a tenfold increase in energy over the previous point. The greatest earthquake ever recorded, which took place in Alaska in 1964, measured over 9 on the Richter scale.

The scale was devised in 1935 by Charles Richter, a US seismologist.

ridge of high pressure elongated area of high atmospheric pressure extending from an anticyclone. On a synoptic weather chart it is shown as a pattern of lengthened isobars. The weather under a ridge of high pressure is the same as that under an anticyclone.

The UK is often under the influence of a ridge of high pressure – in summer it originates from the Azores, and in winter from Scandinavia.

rift valley valley formed by the subsidence of a block of the Earth's ◊crust between two or more parallel ◊faults. Rift valleys are steep-sided and form where the crust is being pulled apart, as at ◊ocean ridges, or in the Great Rift Valley of E Africa.

river long water course that flows down a slope along a channel. It originates at a point called its ***source***, and enters a sea or lake at its ***mouth***. Along its length it may be joined by smaller rivers called ***tributaries***. A river and its tributaries are contained within a ◊drainage basin.

rift valley

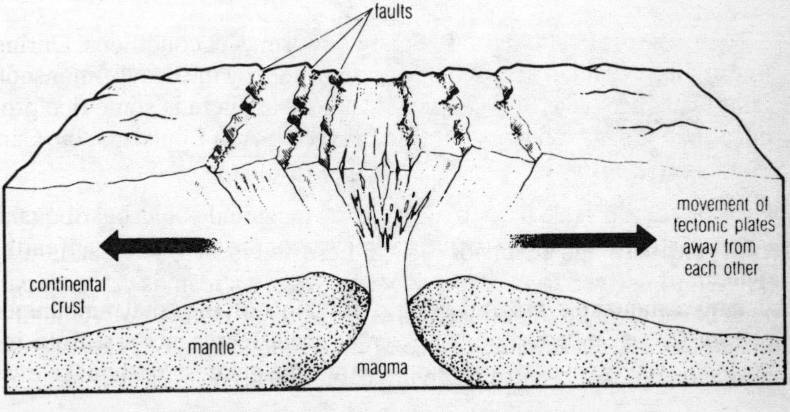

river

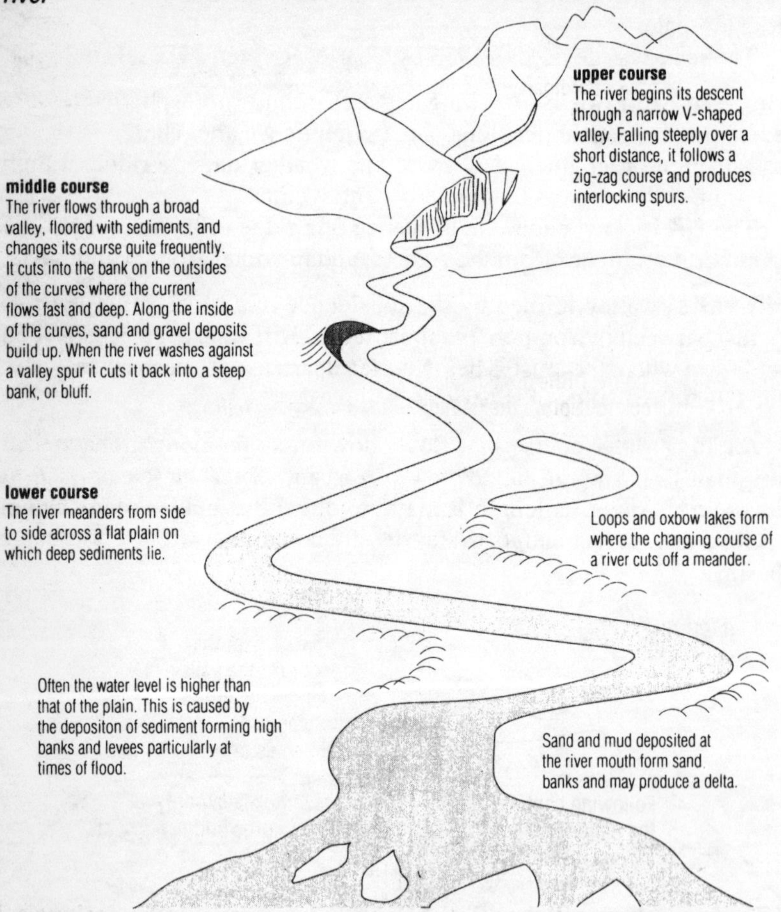

river capture the diversion of the headwaters of one river into a neighbouring river. River capture occurs when a stream is carrying out rapid ◊headward erosion (backwards erosion at its source). Eventually the stream will cut into the course of a neighbouring river, causing the headwaters of that river to be diverted, or 'captured'.

river capture

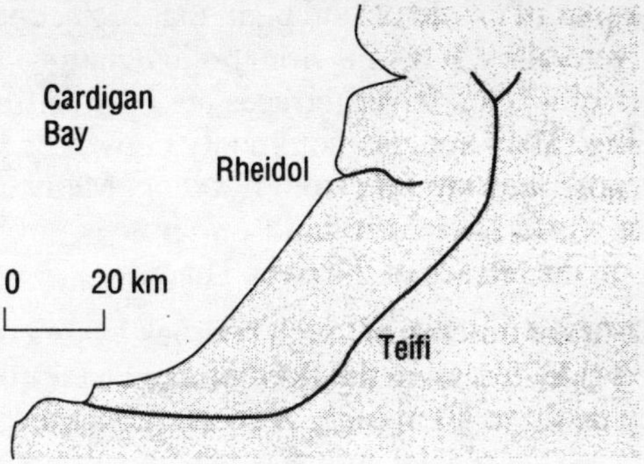

The river Rheidol cuts back along a band of weaker rock to capture the headquarters of the river Teifi.

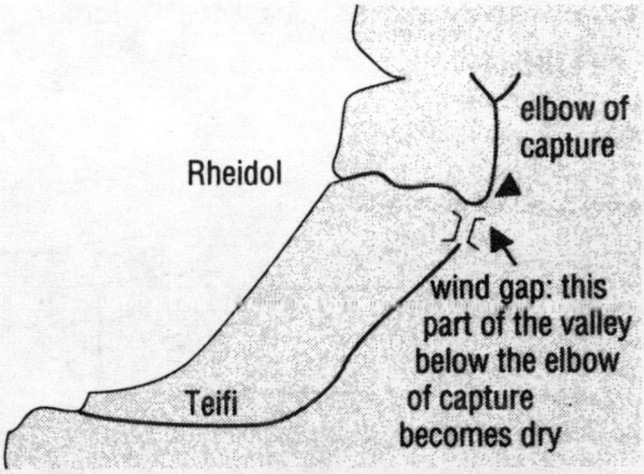

Following capture, rejuvenation has taken place along the upper Rheidol. This has resulted in a gorge and incised meanders.

The headwaters will then flow down to a lower level (often making a sharp bend, called an elbow of capture) over a steep slope, called a ◊knickpoint. A waterfall will form here. ◊Rejuvenation then occurs, causing rapid downwards erosion.

river cliff or ***bluff*** steep slope forming the outer bank of a ◊meander (bend in a river). It is formed by the undercutting of the river current, which at its fastest when it sweeps around the outside of the meander.

river terrace part of an old ◊flood plain that has been left perched on the side of a river valley. It results from ◊rejuvenation, a renewal in the erosive powers of a river. River terraces are very fertile and are often used for farming. They are also commonly chosen as sites for human settlement because they are safe from flooding. Many towns and cities throughout the world have been built on terraces, including London, which is built on the terrace of the river Thames.

roche moutonnée outcrop of tough bedrock having one smooth side and one jagged side, found on the floor of a ◊glacial trough (U-shaped valley). It may be up to 40 m high. A roche moutonnée is a feature of glacial erosion – as a glacier moved over its surface, ice and debris eroded its upstream side by corrasion, smoothing it and creating long scratches or striations. On the sheltered downstream side fragments of rock were plucked away by the ice (see ◊plucking), causing it to become steep and jagged.

roche moutonnée

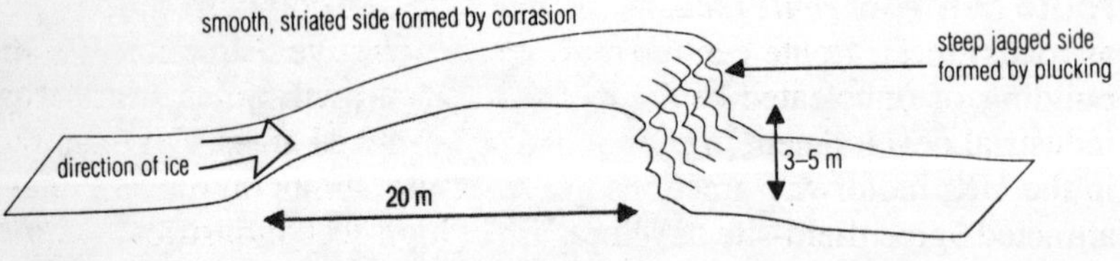

rock hard solid mass composed of mineral particles that have become cemented together. Rocks may be classified into three main groups: ◊igneous, ◊sedimentary, and ◊metamorphic rocks.

Igneous rock is made from magma (molten rock) that has solidified on or beneath the Earth's surface – for example, basalt and granite; ***sedimentary rock*** is formed by the compression of deposited particles – for example, sandstone from sand particles, limestone from the remains of sea creatures; ***metamorphic rock*** is formed by changes in existing igneous or sedimentary rocks under high pressure or temperature, or chemical action – for example, marble is formed from limestone.

roll-on/roll-off or ***ro-ro*** method of loading and unloading cargo ships. Lorries drive straight on board a specially designed vessel and then drive off at the ship's destination without unloading their cargo. This saves time and money, although valuable cargo space is taken up by the lorries and special ◊port facilities are needed. Ro-ro is best suited to shorter journeys where handling costs are proportionately higher. Cross-Channel traffic between the UK and France and Belgium is an example.

root crop plant cultivated for its swollen edible root (which may or may not be a true root). Potatoes are the major temperate root crop; the major tropical root crops are cassava, yams, and sweet potatoes. Together they are second in importance only to cereals as human food. Roots have a high carbohydrate content, but their protein content rarely exceeds 2%. Consequently, communities relying almost exclusively upon roots may suffer from protein deficiency (see ◊malnutrition).

Root crops are also used as animal feed, and may be processed to produce starch, glue, and alcohol. Sugar beet has largely replaced sugar cane as a source of sugar in Europe.

rotational bush fallowing type of ◊shifting cultivation.

route centre or ***route focus*** point where lines of communication (such as roads) meet. Route centres may act as attractive ◊situations for the building of ◊nucleated settlements, or as ◊growth poles stimulating industrial development in economically advanced areas. For example, in the UK, motorway junctions and crossing points on the M25 have attracted ◊greenfield-site development of high-tech industries.

runoff the amount of water that flows off the land, either through streams or over the surface.

rural depopulation loss of people from remote country areas to cities; it is an effect of ◊migration. In poor countries, large-scale migration to urban core regions (see ◊core and periphery) may deplete the countryside of resources and workers. The population left behind will be increasingly aged, and agriculture declines.

In parts of the UK (for example, East Anglia), mechanization of agriculture in the 1950s and 1960s has led to the ◊push factor of declining job prospects, and especially the young have moved away. Rural depopulation may result in reduced village services.

S

Sahel (Arabic *sahil* 'coast') marginal area to the south of the Sahara, from Senegal to Somalia, where the desert is gradually encroaching. The desertification is partly due to climatic change but has also been caused by the pressures of a rapidly expanding population, which have led to overgrazing and the destruction of trees and scrub for fuelwood. In recent years many famines have taken place in the area.

The average rainfall is about 500 mm a year.

salinization the accumulation of salt in water or soil. Water may be desalinized to make it drinkable.

saltation the bouncing of rock particles along a river bed. It is the means by which ◊bedload (material that is too heavy to be carried in suspension) is transported downstream.

salt marsh wetland with characteristic vegetation that is tolerant of sea water. Salt marshes develop around ◊estuaries and on the sheltered side of sand and shingle ◊spits. They usually have a network of creeks and drainage channels by which tidal waters enter and leave the marsh.

San Andreas Fault fault stretching for 1,125 km in a NW–SE direction through California, USA. It marks a conservative margin, where two plates slide past each other. The fault poses a serious threat to the nearby cities of Los Angeles and San Francisco because the friction created by the plates' movement gives rise to frequent, destructive earthquakes. For example, in 1906 an earthquake originating from the fault almost destroyed San Francisco and killed over 3,000 people.

The San Andreas fault has mountains on either side, formed by the movements of the plates; rivers flow along its course because it represents a line of weakness that can be easily eroded.

sand loose grains of rock, sized 0.02–2.00 mm in diameter, consisting chiefly of ◊quartz, but owing their varying colour to mixtures of other minerals. Sand is used in cement-making, as an abrasive, in glass-making, and for other purposes.

Sands may eventually consolidate to form the sedimentary rock ◊sandstone.

sandstone ◊sedimentary rock formed from the consolidation of sand. Sandstones are commonly permeable and porous, and may form freshwater ◊aquifers. They are mainly used as building materials.

satellite town new town planned and built to serve a particular local industry, or as a dormitory or overspill town for people who work in nearby urban areas. Satellite towns in Britain include Port Sunlight near Birkenhead in Cheshire, built to house workers at Lever Brothers soap factories. Problems once caused by lack of social amenities for early arrivals are now overcome by basing a new town on an existing centre.

savanna extensive open tropical grasslands, with scattered trees and shrubs. Savannas cover large areas of Africa, North and South America, and N Australia.

scale on a map, the distance between two places compared to the distance in the real world. The scale may be shown by a representative ***fraction*** (such as 1:50,000 or 1:25,000); a ***scale line***; or a ***statement*** (such as 'Two centimetres equal one kilometre'). All maps should indicate their scale.

Scales vary considerably depending on the purpose of the map. For example, 1:10,000-series Ordnance Survey maps are suited to local studies; a typical road atlas will use a scale of 1:200,000; and a map of the world may use a scale of 1:80,000,000.

The word 'scale' also refers to the size of an area or process that is being discussed. Many geographical principles – such as migration, climate, and pollution – operate on several scales: local, regional, national, international, and global. It is important to study examples at many different scales.

scarp and dip the two slopes of an ◊escarpment. The scarp is usually steep, while the dip slopes gently.

scatter diagram graph in which the value of one set of information is plotted against that of another as a series of points; for example, life expectancy against gross national product (GNP). The points are then examined to see if they show any underlying trend by means of a ***best-fit line***. This is a straight line drawn so that its distance from the various points is as short as possible.

Scatter diagrams are a first step to seeing whether any connection, or ◊correlation, exists between two sets of data. The closer the points are to forming a perfect line, the greater the degree of correlation between the two data sets.

science park site near a university on which high-technology industrial businesses are housed, so that they can benefit from the research expertise of the university's scientists. Science parks originated in the USA in the 1950s.

By 1985 the UK had 13 science parks, the first being Heriot-Watt in Edinburgh and the most successful in Cambridge.

scree pile of rubble and sediment that collects at the foot of a mountain range or cliff. The rock fragments that form scree are usually broken off by the action of frost (◊freeze-thaw weathering).

sea breeze gentle coastal wind blowing off the sea towards the land. It is most noticeable in summer when the warm land surface heats the air above it and causes it to rise. Cooler air from the sea is drawn in to replace the rising air, so causing an onshore breeze. At night and in winter, air may move in the opposite direction, forming a ◊land breeze.

seafloor spreading growth of the ocean ◊crust outwards (sideways) from ocean ridges. The concept of seafloor spreading has been combined with that of continental drift and incorporated into ◊plate tectonics.

season period of the year having a characteristic climate. The change in seasons is mainly due to the change in attitude of the Earth's axis in relation to the Sun, and hence the position of the Sun in the sky at a particular place. In temperate latitudes four seasons are recognized: spring, summer, autumn (fall), and winter. Tropical regions have two seasons – the wet and the dry. Monsoon areas around the Indian Ocean have three seasons: the cold, the hot, and the rainy.

sea breeze

daytime sea breeze

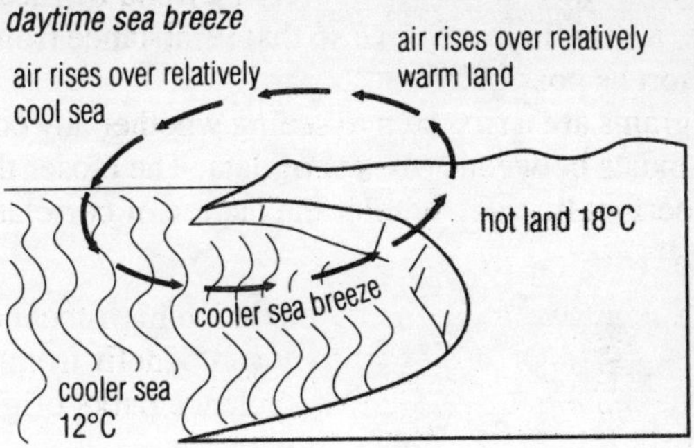

night-time land breeze

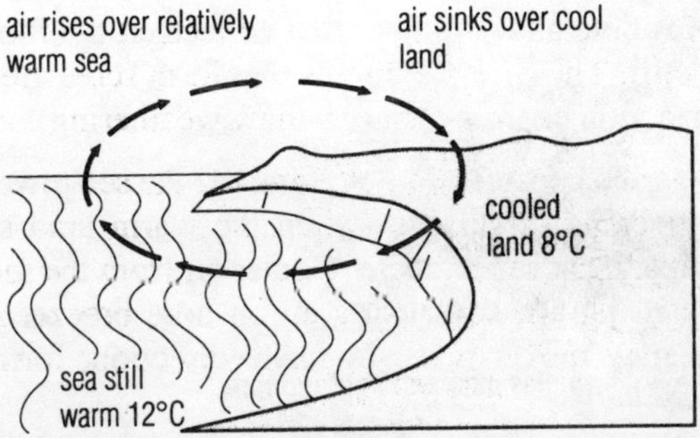

secondary industry manufacturing from raw materials; see ◊industrial sector.

sector theory ◊model of urban land use in which the various land-use zones are shaped like wedges radiating from the central business district. According to sector theory, the highest prices for land are found along transport routes (especially roads), and once an area has gained a reputation for a particular type of land use (such as industry), it will attract the same land users as the city expands outwards.

The theory was suggested by H Hoyt in 1939 from his studies of Chicago, USA.

sector theory

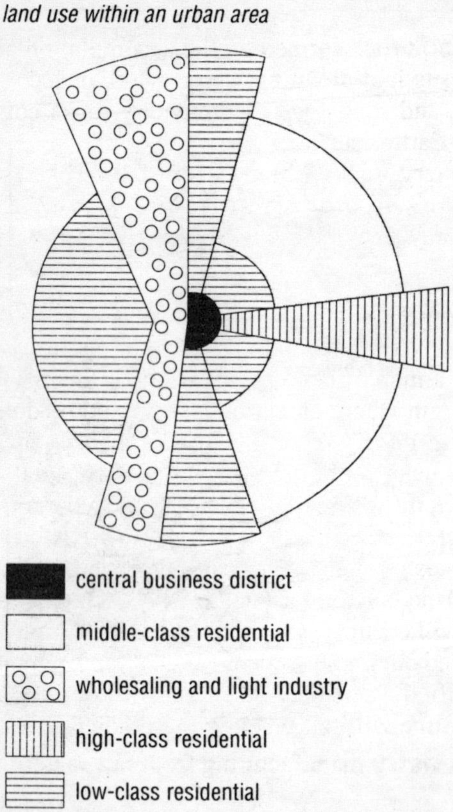

sedentary agriculture ◊farming system in which the farmer remains settled in one place. It contrasts with ◊shifting cultivation and ◊nomadic pastoralism.

sediment any loose material that has 'settled' – become deposited from suspension in water, ice, or air, generally as the water-current or wind velocity decreases. Typical sediments are, in order of increasing particle size: clay, mud, silt, sand, gravel, pebbles, cobbles, and boulders.

Accumulations of sediments containing valuable metals are called ◊placer deposits.

sedimentary rock rock formed by the accumulation and cementing together of deposits laid down by water, wind, or ice – for example, limestone, shale, and sandstone. Sedimentary rocks cover more than two-thirds of the Earth's surface.

Most sedimentary rocks show distinct layering (stratification), caused by alterations in composition or by changes in rock type. These strata, or beds, may become folded or faulted by the movement of tectonic plates.

seismic-gap theory recent theory that aims to predict the location of ◊earthquakes. When records of past earthquakes are studied and plotted onto a map, it becomes possible to identify areas along a fault or plate margin where an earthquake should be due – such areas are called ***seismic gaps***. According to the theory, an area that has not had an earthquake for some time will have a great deal of stress building up, which must eventually be released in the form of an earthquake.

Although the seismic-gap theory can suggest areas that are likely to experience an earthquake, it does not enable scientists to predict when that earthquake will occur.

Research carried out in the vicinity of the ◊San Andreas fault in California has identified a seismic gap at the town of Parkfield, near San Francisco. Only time will tell whether the prediction will prove to be correct.

seismic wave energy wave generated by an ◊earthquake; it is responsible for the shaking of the ground and the conversion of soft deposits, such as clay, to a jellylike state (◊liquefaction).

The different speeds at which seismic waves move through the Earth have been used by scientists to deduce the Earth's layered internal structure.

seismogram or ***seismic record*** trace, or graph, of an earthquake's activity over time, recorded by a seismograph. It is used to determine the magnitude and duration of an earthquake.

seismograph

a seismogram recorded by a seismograph

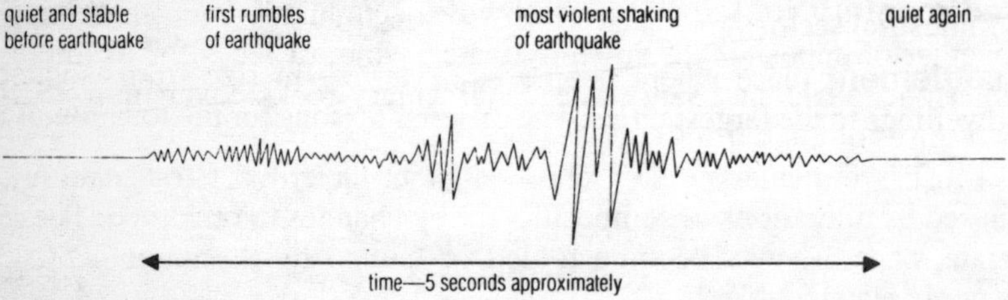

seismograph instrument used to record the activity of an ◊earthquake. A heavy inert weight is suspended by a spring and attached to this is a pen that is in contact with paper on a rotating drum. During an earthquake the instrument frame and drum move, causing the pen to record a zigzag line on the paper; the pen does not move.

seismology the study of ◊earthquakes.

self-help project any scheme for a community to help itself under official guidance. The most popular self-help projects in the developing world are aimed at improving conditions in ◊shanty towns. Organized building lots are commonly provided, together with properly laid-out drains, water supplies, roads, and lighting. ◊Squatters are expected to build their own homes on the prepared sites, perhaps with loans provided by the government or other agencies. An example is the Arumbakkam scheme in Madras, India, begun 1977. Alternatively, 'basic shell' housing may be provided, as in parts of São Paulo, Brazil, and Colombia.

selva equatorial rainforest, such as that in the Amazon basin in South America.

service centre another name for a ◊central place.

service industry commercial activity that provides and charges for various services to customers (as opposed to manufacturing or supplying goods), such as restaurants, the tourist industry, cleaning, hotels, and the retail trade (shops and supermarkets).

With the decline in the manufacturing sector in many Western countries, service industries have become major employers of labour.

Service industries may also be described as tertiary industries (see ◊industrial sector).

settlement place where people live, varying in size from isolated dwellings to the largest cities. The original reasons for the location of a

settlement

types of settlement

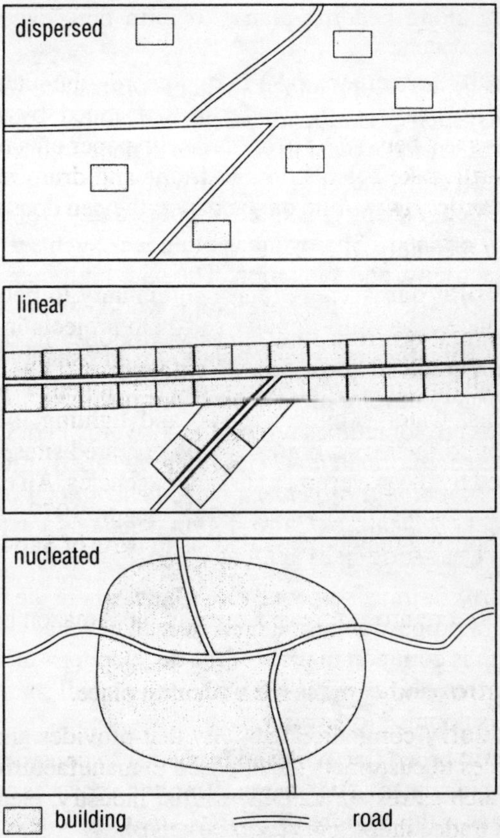

settlement are often found by examining the ◊site – for example, it may be well supplied with water or easily defended from enemies.

Settlements may take many different forms – for example, ***dispersed settlements*** are made up of isolated buildings scattered over a wide area; ***linear developments*** are long and ribbonlike, and usually develop along roads; ***nucleated settlements*** are grouped around a central point, or nucleus, such as a spring or a crossroads.

shale fine-grained and finely-layered ◊sedimentary rock composed of silt and clay, usually formed in lowland areas. Shale is a weak rock, splitting easily along bedding planes to form thin, even slabs. It is impermeable to water.

shanty town group of unplanned shelters constructed from cheap or waste materials (such as cardboard, wood, and cloth). Shanty towns are commonly located on the outskirts of cities in poor countries, or within large cities on derelict land or near rubbish tips.

Land available for shanties is often of poor quality (for example, too steep or poorly drained). Shanty areas often lack such services as running water, electricity, and sanitation. They are high-density developments; that is, crowded. In the developing world, shanty towns result from mass ◊migration from rural areas in response to ◊pull factors, especially the perceived prospects of employment. One solution to uncontrolled shanty development is ◊self-help projects.

Shanty towns are sometimes referred to by words in or derived from the local language; in Brazil, *favela*; in India, especially around Calcutta, *basti*; in Peru, *colonia proletaris*; in Tunisia, *gourbi* or *bidonville*; and in Venezuela, *rancho*. In Cairo, Egypt, squatters have taken over the City of the Dead, originally built as a burial ground.

sharecropping farming someone else's land, where the farmer gives the landowner a proportion of the crop instead of money. This system of rent payment is common in parts of the developing world; for example, in India. Often the farmer is left with such a small share of the crop that he or she is doomed to poverty.

shield volcano broad flat ◊volcano formed at a ◊constructive margin between tectonic plates. The magma (molten rock) associated with shield volcanoes is thin and free-flowing. An example is Mauna Loa in

Hawaii. A ◊composite volcano, on the other hand, is formed at a destructive margin.

shifting cultivation ◊farming system where farmers move on from one place to another. The most common form is ***slash-and-burn*** agriculture: land is cleared by burning, so that crops can be grown. After a few years, soil fertility is reduced and the land is abandoned. A new area is cleared while the old land recovers its fertility.

Slash-and-burn is practised in many tropical forest areas, such as the Amazon region, where yams, manioc, and sweet potatoes can be grown. This system works well while population levels are low, but where there is ◊overpopulation, the old land will be reused before soil fertility has been restored. A variation of this system, found in parts of Africa, is rotational bush fallowing.

Shimbel index a mathematical measurement of the ◊accessibility of a transport network. It is the total of the number of links or edges that form the shortest path between each ◊node (junction or place) and all other nodes in the network.

sial the substance of the Earth's continental ◊crust, as distinct from the ◊sima of the ocean crust. The name is derived from ***si***lica and ***al***umina, its two main chemical constituents. Sial is often rich in granite.

silage fodder preserved through controlled fermentation in a ◊silo, an airtight structure that presses green crops. It is used as a winter feed for livestock.

Silicon Glen area in central Scotland, around Glenrothes ◊new town, where there are many electronics firms. By 1986 Glenrothes had over 21% of its workforce employed in electrical engineering, especially high-tech firms. Many of the firms here are owned by US and other foreign companies.

Silicon Valley region in S California, USA, with a high concentration of high-tech industries connected with microchip production. In the UK, there is a similar concentration along the 'western corridor', following the route of the M4 between London and Bristol, and in ◊Silicon Glen.

sill sheet of igneous rock created by the intrusion of magma (molten rock) between layers of pre-existing rock. (A ◊dyke, by contrast, is formed when magma cuts *across* layers of rock.) An example of a sill in the UK is the Great Whin Sill, which forms the ridge along which Hadrian's Wall was built.

The rock of which sill is formed is called ***dolerite***. It is extremely resistant to erosion and weathering, and often forms ridges in the landscape or cuts across rivers to create ◊waterfalls.

silo in farming, an airtight tower in which ◊silage is made by the fermentation of freshly cut grass and other forage crops.

sima the substance of the Earth's oceanic ◊crust, as distinct from the ◊sial of the continental crust. The name is derived from ***si***lica and ***ma***gnesium, its two main chemical constituents.

sink hole funnel-shaped hollow in an area of limestone. A sink hole is usually formed by the enlargement of a joint, or crack, by ◊solution (the dissolving effect of water). It should not be confused with a ◊swallow hole, or swallet, which is the opening through which a stream disappears underground when it passes onto limestone.

sirocco hot, normally dry and dust-laden wind that blows from the deserts of N Africa across the Mediterranean into S Europe. It occurs mainly in the spring.

site the land on which a settlement is built, described according to its own characteristics – for example, its altitude, drainage, proximity to reliable water supplies, building materials, and the availability of farmland. By contrast, the ◊situation of a settlement is described in relation to its surroundings.

All settlements, however large, started from an initial point and may therefore be classified according to the original characteristics of that site – for example, London's site may be described as an area of slightly raised, dry ground at a convenient crossing point of the river Thames.

site of special scientific interest (SSSI) in the UK, area designated as being of particular environmental interest by one of the regional bodies of the government's Nature Conservancy Council.

Numbers fluctuate, but there were over 5,000 SSSIs in 1991, covering about 6% of Britain. Although SSSIs enjoy some legal protection, this does not in practice always prevent damage or destruction; during 1989, for example, 44 SSSIs were so badly damaged that they were no longer worth protecting.

situation the relationship of a settlement to other places and its surrounding area; for example, its ◊topography and routeways. The situation of a settlement differs from its ◊site in that site refers only to the land on which the settlement is built whereas situation links the land to its surrounding physical and human features.

slash-and-burn common form of ◊shifting cultivation, whereby natural vegetation is cut and burned, and the clearing then farmed for a few years until the soil loses its fertility, whereupon farmers move on and leave the area to regrow.

slate fine-grained, usually grey ◊metamorphic rock that splits readily into thin slabs. It is the metamorphic equivalent of ◊shale.

Slate is highly resistant to erosion and weathering and often forms upland areas, such as those in North Wales. It main use is as a roofing material.

sleet precipitation consisting of a mixture of water and ice, formed from melted falling snow or hail. In North America the term refers to precipitation in the form of ice pellets smaller than 5 mm.

slip-off slope gentle slope forming the inner bank of a ◊meander (bend in a river). It is formed by the deposition of fine silt, or alluvium, by slow-flowing water.

As water passes round a meander the fastest current sweeps past the outer bank, eroding it to form a steep river cliff. Water flows more slowly past the inner bank, and as it reduces speed the material it carries is deposited around the bank to form a slip-off slope.

slum area of poor-quality housing. Slums are typically found in parts of the ◊inner city in rich countries and in older parts of cities in poor countries. Slum housing is usually densely populated, in a bad state of repair, and has inadequate services (poor sanitation, for example). Its occupants are often poor with low rates of literacy.

The clearing of these areas has often been a priority in urban redevelopment (in the UK, for example, the Glasgow Eastern Area Redevelopment scheme of the 1960s). Where slum housing is structurally sound, ◊urban renewal (renovation of old housing and improvement of services) may be more appropriate, as in Glasgow in the 1970s.

slurry form of manure composed mainly of liquids. Slurry is collected and stored on many farms, especially when large numbers of animals are kept in factory units (see ◊factory farming). When slurry tanks are accidentally or deliberately breached, large amounts can spill into rivers, killing fish and causing ◊eutrophication.

Slurry spills in the UK increased enormously in the 1980s and were by 1991 running at rates of over 4,000 a year. Tighter regulations were being introduced to curb this pollution.

smog natural fog containing impurities (unburned carbon and sulphur dioxide) from domestic fires, industrial furnaces, certain power stations, and internal-combustion engines (petrol or diesel). It can cause substantial illness and loss of life, particularly among chronic bronchitics, and damage to wildlife.

The London smog of 1952 killed 4,000 people from heart and lung diseases, probably caused by inhalation of sulphuric acid droplets. The use of smokeless fuels, the treatment of effluent, and penalties for excessive smoke from poorly maintained and operated vehicles can be extremely effective in cutting down smog, as in London, but it still occurs in many cities.

snout the front end of a ◊glacier, representing the furthest advance of the ice at any one time. Deep cracks, or ◊crevasses, and ice falls are common.

Because the snout is the lowest point of a glacier it tends to be affected by the warmest weather. Considerable melting takes place, and so it is here that much of the rocky material carried by the glacier becomes deposited. Material dumped just ahead of the snout may form a ◊terminal moraine.

The advance or retreat of the snout depends upon the glacier budget – the balance between ◊accumulation (the addition of snow and ice to the glacier) and ◊ablation (their loss by melting and evaporation).

snow precipitation in the form of soft, white, crystalline flakes caused by the condensation in air of excess water vapour below freezing point. Light reflecting in the crystals, which have a basic hexagonal (six-sided) geometry, gives snow its white appearance.

social fabric the make-up of an area in terms of its social geography, such as class, ethnic composition, employment, education, and values.

soil loose covering of broken rocky material and decaying organic matter overlying the bedrock of the Earth's surface. Various types of soil develop under different conditions: deep soils form in warm wet climates and in valleys; shallow soils form in cool dry areas and on slopes.

Soils influence the type of agriculture employed in a particular region – light well-drained soils favour arable farming, whereas heavy clay soils give rise to lush pasture land.

soil creep gradual movement of soil down a slope. As each soil particle is dislodged by a raindrop it moves slightly further downhill. This eventually results in a mass downward movement of soil on the slope.

Evidence of soil creep includes the formation of terracettes (steplike ridges along the hillside), leaning walls and telegraph poles, and trees that grow in a curve to counteract progressive leaning.

soil erosion the wearing away and redistribution of the Earth's soil layer. It is caused by the action of water, wind, and ice, and also by improper methods of agriculture. If unchecked, soil erosion results in the formation of deserts (see ◊desertification). It has been estimated that 20% of the world's cultivated topsoil was lost between 1950 and 1990.

If the rate of erosion exceeds the rate of soil formation (from rock), then the land will decline and eventually become infertile.

The removal of forests (◊deforestation) or other vegetation often leads to serious soil erosion, because plant roots bind soil, and without them the soil is free to wash or blow away, as in the American ◊dust bowl. The effect is worse on hillsides, and there has been devastating loss of soil where forests have been cleared from mountainsides, as in

Madagscar. Improved agricultural practices such as contour ploughing are needed to combat soil erosion. Windbreaks, such as hedges or strips planted with coarse grass, are valuable, and organic farming can reduce soil erosion by as much as 75%.

solar energy energy derived from the Sun's radiation. The amount of energy falling on just 1 sq km is about 4,000 megawatts, enough to heat and light a small town. This energy is not easy to harness, but a number of solar-energy technologies have emerged during recent years.

Solar heaters have industrial or domestic uses. They usually consist of a black (heat-absorbing) panel containing pipes through which air or water, heated by the Sun, is circulated, either by thermal ◊convection or by a pump. Solar energy may also be harnessed indirectly using *solar cells* (photovoltaic cells) made of panels that generate electricity when illuminated by sunlight.

Although it is difficult to generate a high output from solar energy compared to sources such as nuclear or fossil fuels, it is a major nonpolluting and renewable energy source and is used as far north as Scandinavia as well as in the southwest USA and in Mediterranean countries.

solar pond natural or artificial 'pond', such as the Dead Sea, in which salt becomes more soluble in the Sun's heat. Water at the bottom becomes saltier and hotter, and is insulated by the less salty water layer at the top. Temperatures at the bottom reach about 100°C and can be used to generate electricity.

solar radiation radiation given off by the Sun, consisting mainly of visible light, ultraviolet radiation, and infrared radiation, although the whole spectrum of electromagnetic waves is present, from radio waves to X-rays. High-energy charged particles such as electrons are also emitted, especially from solar flares. When these reach the Earth, they cause magnetic storms (disruptions of the Earth's magnetic field), which interfere with radio communications.

solifluction the downhill movement of topsoil that has become saturated with water. Solifluction is common in periglacial environments (those bordering glacial areas) during the summer months, when the frozen topsoil melts to form an unstable soggy mass. This may then flow slowly downhill under gravity to form a ***solifluction lobe*** (a tonguelike feature).

Solifluction material, or head, is found at the bottom of chalk valleys in southern England; it is partly responsible for the rolling landscape typical of chalk scenery.

solstice either of the days on which the Sun is farthest north or south of the celestial equator each year. The ***summer solstice***, when the Sun is farthest north, occurs around June 21; the ***winter solstice*** around Dec 22.

solution the dissolving in water of minerals within a rock. It may result in weathering (for example, when weakly acidic rainfall causes ◊carbonation) and erosion (when flowing water passes over rocks).

Solution commonly affects limestone and chalk, and may be responsible for forming features such as ◊sink holes.

spa town town with a spring, the water of which, it is claimed, has the power to cure illness and restore health. Spa treatment involves drinking and bathing in the naturally mineralized spring water.

Spa towns in the UK include Harrogate, Tunbridge Wells, Epsom, Bath, Leamington, and Llandrindod Wells.

Spearman's rank correlation coefficient index of the strength of the relationship between two sets of information; for example, altitude and temperature. The value of the index varies between +1 (a perfect positive correlation) and –1 (a perfect negative correlation); 0 = no correlation.

The formula for the calculation of the Spearman's rank correlation coefficient R is:

$$R = \frac{1 - (6 - d^2)}{(n^3 - n)}$$

where n is the number of pairs of data and d is the difference in rank.

As a first step to investigate correlation, a ◊scatter diagram can be plotted, to see how near to a straight line the points lie.

sphere of influence the area surrounding a settlement that is affected by the settlement's activities. The sphere of influence may be mapped by looking at the catchment areas of various services or by considering local-newspaper circulation, delivery areas, and public-transport destinations.

All settlements are ◊central places with functions such as housing, administration, health care, shopping, and recreation. The number and hierarchy of these functions determines the area of the sphere of influence and depend on the size and importance of the settlement – whether it is a village, town, or city.

spit ridge of sand or shingle projecting from the land into a body of water. It is deposited by a current carrying material from one direction to another across the mouth of an inlet (◊longshore drift). Deposition in the brackish water behind a spit may result in the formation of a ◊salt marsh.

spit

Blakeney spit, Norfolk

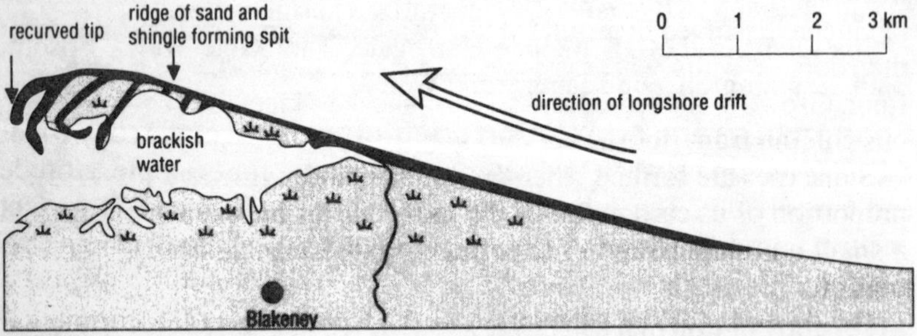

spring a natural flow of water from the ground, formed where the water table meets the ground's surface. The source of water is rain that has fallen on the overlying rocks and percolated through. During its passage the water may have dissolved mineral substances, which may then be precipitated at the spring.

Springs provide a reliable source of water and are therefore attractive sites for settlements (for example, ◊spa towns).

spring line geological feature where water springs up in several places along the edge of a permeable rock escarpment.

spur ridge of rock jutting out into a valley or plain. In mountainous areas rivers often flow around ◊interlocking spurs because they are not powerful enough to erode through the spurs. Spurs may be eroded away by large and powerful glaciers to form ◊truncated spurs.

squatter illegal occupant of a building or land. People in ◊shanty towns are often squatters. In many ◊inner-city areas in rich countries, homeless people squat in derelict or unoccupied buildings.

A distinction is made between squatters and slum dwellers, since slum dwellers are legal occupants of substandard buildings.

stack isolated pillar of rock that has become separated from a headland by ◊coastal erosion. It is usually formed by the collapse of an ◊arch. Examples of stacks in the UK are the Needles, off the Isle of Wight, which are formed of chalk.

A stack is usually only a metre or so across but is by definition tall enough to be visible always above the water. Further erosion will reduce it to a ◊stump, which is exposed only at low tide.

stalactite and stalagmite cave structures formed by the deposition of calcite dissolved in ground water. ***Stalactites*** grow downwards from the roofs or walls whereas ***stalagmites*** grow upwards from the cave floor. Growing stalactites and stalagmites may meet to form a continuous column from floor to ceiling.

Stalactites are formed when ground water, hanging as a drip, loses a proportion of its carbon dioxide into the air of the cave. This results in a small amount of calcite being deposited. Successive drips build up the stalactite over many years.

steel alloy of iron that can be produced to varying degrees of hardness. It has innumerable uses, including ship and car manufacture, skyscraper frames, and machinery and precision tools of all kinds. The USA, Russia, Ukraine, and Japan are the main steel producers.

Steel is produced by removing impurities, such as carbon, from raw, or pig, iron produced by a ◊blast furnace. The steel produced is cast into ingots, which can be worked when hot by hammering (forging) or pressing between rollers to produce sheet steel.

The last 50 years have seen a decline in the number of steel works in the UK. In 1992 British Steel had five main plants – at Port Talbot and Llanwern in S Wales; Ravenscraig in Strathclyde, Scotland; and Teesside and Scunthorpe in NE England. These works now import most of their raw materials. Many steelworks have remained in their original locations even though the source of raw materials has changed (see ◊industrial inertia).

steppe the temperate grasslands of Europe and Asia. Arable and pastoral farming are carried out there.

Stevenson screen box designed to house weather-measuring instruments such as thermometers. It is kept off the ground by legs, has louvred sides to encourage the free passage of air, and is painted white to reflect heat radiation, since what is measured is the temperature of the air, not of the sunshine.

stolport (abbreviation for ***s***hort ***t***ake***o***ff and ***l***anding ***port***) airport that can be used by planes adapted to a shorter than normal runway. Such planes tend to have a restricted flying range. Stolport sites are found in built-up areas (such as London Docklands) where ordinary planes would not be able to land safely.

storm surge abnormally high tide brought about by a combination of a severe atmospheric depression (very low pressure) over a shallow sea area, high spring tides, and winds blowing from the appropriate direction. A storm surge can cause severe flooding of lowland coastal regions and river estuaries.

Bangladesh is particularly prone to surges, being sited on a low-lying ◊delta where the Indian Ocean funnels into the Bay of Bengal. In May 1991, 125,000 people were killed there in such a disaster. In Feb 1953 more than 2,000 died when a North Sea surge struck the Dutch and English coasts.

strata (singular ***stratum***) layers or ◊beds of ◊sedimentary rock.

striation scratch formed by the movement of a glacier over a rock surface. Striations are caused by the scraping of rocky debris embedded in the base of the glacier (◊corrasion), and provide an useful indicator of the direction of ice flow in past ◊ice ages. They are common features of ◊roche moutonnées.

stump low outcrop of rock formed by the erosion of a coastal ◊stack. Unlike a stack, which is exposed at all times, a stump is exposed only at low tide. Eventually it will be worn away completely, leaving a ◊wave-cut platform.

subduction zone in plate tectonics, a region in which one plate descends below another. Subduction zones are a feature of ◊destructive margins (regions where plates collide); most are marked by ocean trenches.

subglacial beneath a glacier. Subglacial rivers are those that flow under a glacier; subglacial material is debris that has been deposited beneath glacier ice. Features formed subglacially include ◊drumlins and ◊eskers.

subsistence farming farming when the produce is enough to feed only the farmer and family and there is no surplus to sell.

It is the main farming system throughout much of the ◊developing world.

suburb outer part of an urban area. Suburbs generally consist of residential housing and shops of a low order (newsagent, small supermarket; see ◊hierarchy), which act as ◊central places for the local community. Often, suburbs are the most recent growth of an urban area, and their end marks the ◊urban fringe. Their growth may result in ◊urban sprawl.

sugar beet root crop from which sugar is refined. Refineries are located near areas where the crop is grown, since it loses much weight during processing. Only a small percentage of refined sugar in the UK is now produced from sugar cane.

sunshine recorder device for recording the hours of sunlight during a day. The ***Campbell-Stokes sunshine recorder*** consists of a glass sphere that focuses the sun's rays on a graduated paper strip. A track is burned along the strip corresponding to the time that the Sun is shining.

supermarket large self-service shop selling food and household goods. Larger versions, called ◊hypermarkets, are usually situated on the outskirts of cities and towns.

Supermarkets have a high turnover and are therefore able to buy goods in bulk. This cuts down the unit cost and, in turn, the price, which further encourages custom. Cut-price supermarkets have in some places led to the closure of small local shops.

supraglacial on top of a glacier. A supraglacial stream flows over the surface of the glacier; supraglacial material collected on top of a glacier may be deposited to form ◊lateral moraines and ◊medial moraines.

surface runoff overland transfer of water after a rainfall. It is the most rapid way in which water reaches a river. The amount of surface runoff increases given (1) heavy and prolonged rainfall, (2) steep gradients, (3) lack of vegetation cover, (4) saturated or frozen soil. A ◊hydrograph can measure the time the runoff takes to reach the river. ◊Throughflow is another way water reaches a river.

surveying the accurate measuring of the Earth's crust, or of land features or buildings. It is used to establish boundaries, and to evaluate the topography for engineering work.

swallet alternative name for a ◊swallow hole.

swallow hole hole, often found in limestone areas, through which a surface stream disappears underground. It will usually lead to an underground network of caves. Gaping Gill in North Yorkshire is an example.

swash the advance of water and sediment up a beach as a ◊wave breaks. Swash plays a significant role in the movement of beach material by ◊longshore drift, and is responsible for throwing shingle and pebbles up a beach to create ridges called ◊berms.

syncline ◊fold in the rocks of the Earth's crust in which the layers, or beds, bulge downwards. (A fold that arches upwards is an anticline.)

synoptic chart weather chart in which symbols are used to represent the weather conditions experienced over an area at a particular time. Synoptic charts appear on television and newspaper forecasts, although the symbols used may differ.

synoptic chart

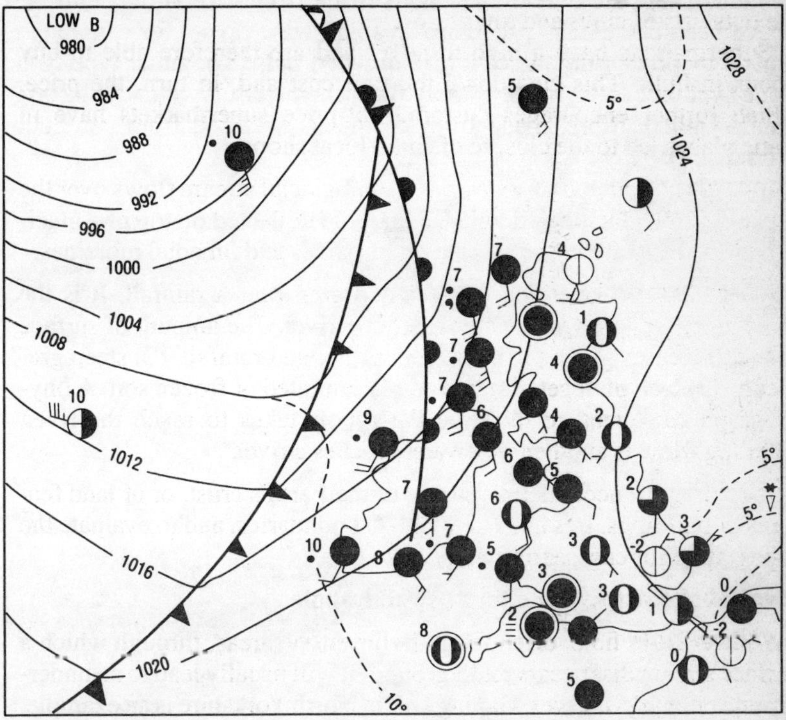

explanation of symbols

cloud amount in oktas (unit used in meteorological reports to describe cloudiness)

0 · 1 or less · 2 · 3 · 4 · 5 · 6 · 7 or more · 8

sky obscured · missing or doubtful data · warm front · cold front

wind speed in knots

calm · 1–2 · 3–7 · 8–12 · 13–17 · for each additional half-feather add 5 knots · 48–52

= mist · ≡ fog · , drizzle · rain and drizzle · • rain · rain and snow · ✱ snow · rain shower

rain and snow shower · snow shower · hail shower · thunderstorm

—— 1004 —— isobars drawn at intervals of 4 millibars · — — 5° — — mean sea surface isotherm

synthetic fibre fibre made by chemical processes, unknown in nature. There are two kinds. One is made from natural materials that have been chemically processed in some way; rayon, for example, is made by processing the cellulose in wood pulp. The other type is the true synthetic fibre, made entirely from chemicals. Nylon was the original synthetic fibre, made from chemicals obtained from petroleum (crude oil).

T

tail in previously glaciated areas, a tapering ridge of debris that has been protected from erosion by a glacier by being on the sheltered side of a lump (◊crag) of more resistant rock.

tariff tax or duty placed on goods when they are imported into a country or trading bloc (such as the European Community) from outside. The aim of tariffs is to reduce imports by making them more expensive.

Tariffs have generally been used by governments to protect home industries from lower-priced foreign goods, and have been opposed by supporters of free trade. For a tariff to be successful, it must not provoke retaliatory tariffs from other countries. Organizations such as the EC, the European Free Trade Association (EFTA), and the General Agreement on Tariffs and Trade (GATT) have worked towards mutual lowering of tariffs between countries.

tenant farming system whereby farmers rent their holdings from a landowner in return for the use of agricultural land.

In 19th-century Britain, most farmland was organized into landed estates containing tenanted farms. A marked change began after World War I when, owing to the agricultural depression, many landowners sold off all or part of their estates, often to the sitting tenant farmers. Although in 1950, 50% of the country's farms were still rented, the current figure is less than 25%.

◊Sharecropping is a variety of tenant farming, used in the the ◊developing world, in which the farmer gives the landowner a proportion of the crop instead of money.

terminal moraine linear, slightly curved ridge of rocky debris deposited at the front end, or snout, of a glacier. It represents the furthest point of advance of a glacier, being formed when deposited material (till), which was pushed ahead of the snout as it advanced, became left behind as the glacier retreated.

A terminal moraine may be a few hundred metres in height; for example, the Franz Joseph glacier in New Zealand has a terminal moraine that is over 400 m high.

terrace farming farming on steep slopes that are terraced to produce a series of flat fields, stepped in appearance. This flat land can then be cultivated. Examples are vineyards in the Rhine valley and paddy rice farming in the Philippines.

tertiary industry service industry; see ◊industrial sector.

textile fabric woven from natural fibres, such as cotton or wool, or from synthetic fibres, such as nylon or polyester. The textile industry is a major secondary, or manufacturing, industry in such countries as China, India, Hong Kong, and South Korea. In the UK the industry is based largely in Yorkshire and Lancashire.

theme park large leisure park with many rides and amusements, linked by a theme or concept. The first was Disneyland in California, USA, opened 1955 to promote the cartoon characters of Walt Disney. Another Disney theme park, EuroDisney, opened in 1992 near Paris.

thermometer instrument for measuring temperature. There are many types, designed to measure different temperature ranges to varying degrees of accuracy. Each makes use of a different physical effect of temperature – for example, the common ***liquid-in-glass thermometers***, such as those containing mercury or alcohol, make use of the expansion of liquids in response to heat.

Third World term originally applied collectively to those countries of Africa, Asia, and Latin America that were not aligned with either the Western bloc (First World) or Communist bloc (Second World). The term later took on economic connotations and was applied to the countries of the ◊developing world.

threshold population the minimum number of people necessary before a particular good or service will be provided in an area. Typically a low-order shop (such as a grocer or newsagent; see ◊hierarchy) may require only 800 or so customers, whereas a higher-order store such as Marks and Spencer may need a threshold of 70,000 to be profitable, and a university 350,000 to be viable.

Thresholds may also be linked to the spending power of customers; this is most obvious in periodic markets in poor countries, where wages are so low that people can buy the goods or services only once in a while.

throughflow the seepage, or percolation, of water through soil. In the hydrological, or water, cycle it is one of the processes responsible for the movement of water from the land to the oceans.

Throughflow is slower than water flow overland but it can be quite rapid, particularly if pipes or gaps exist in the soil (see ⟩pipeflow).

thunderstorm severe storm of very heavy rain, thunder, and lightning.

Thunderstorms are usually caused by the intense heating of the ground surface during summer. The warm air rises rapidly to form tall cumulonimbus clouds with a characteristic anvil-shaped top. Electrical charges accumulate in the clouds and are discharged to the ground as flashes of lightning. Air in the path of lightning becomes heated and expands rapidly, creating shock waves that are heard as a crash or rumble of thunder.

The rough distance between an observer and a lightning flash can be calculated by timing the number of seconds between the flash and the thunder. A gap of 3 seconds represents about a kilometre; 5 seconds represents about a mile.

tidal energy renewable energy derived from the tides. If water is trapped at a high level during high tide, perhaps by means of a barrage across an estuary, it may then be gradually released and its energy exploited to drive turbines and generate electricity.

Several barrage schemes have been proposed for the Bristol Channel, but environmental concerns as well as construction costs have so far prevented any decision from being taken.

tidal power station ⟩hydroelectric power plant that uses the 'head' of water created by the rise and fall of the ocean tides to spin the water turbines. An example is located on the estuary of the river Rance in the Gulf of St Malo, Brittany, France, which has been in use since 1966.

tidal wave misleading name for a ⟩tsunami.

tide rise and fall of sea level due to the gravitational forces of the Moon and Sun. The highest or ***spring tides*** are at or near new and full Moon; the lowest or ***neap tides*** when the Moon is in its first or third quarter. Some seas, such as the Mediterranean, have very small tides.

till or ***boulder clay*** deposit of clay, mud, gravel, and boulders left by a ◊glacier. It is unsorted, with all sizes of fragments mixed up together, and shows no stratification; that is, it does not form clear layers or ◊beds.

Till may be laid down evenly to form a fertile soil, as in East Anglia, or it may be deposited to create landforms characteristic of glacial deposition, such as drumlins and moraines.

tombolo spit, or ridge of sand or shingle, that connects the mainland to an island; for example Chesil bank, which extends 19 km from Abbotsbury in Dorset, England, to the Isle of Portland.

tombolo

Chesil bank joins the Isle of Portland to the mainland in South Dorset

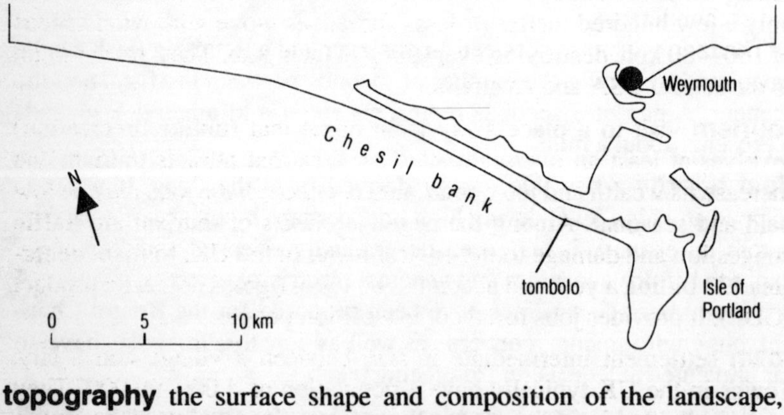

topography the surface shape and composition of the landscape, comprising both natural and artificial features. Topographical features include the relief and contours of the land; the distribution of mountains, valleys, and human settlements; and the patterns of rivers, roads, and railways. Such features are shown on ***topographical maps*** (for example, those produced by the ◊Ordnance Survey).

topological map simplified map of a network, such as a transport network, in which ◊connectivity (the way lines join together) is preserved but shape and size are not. An example is the map of the London Underground network, which clearly displays the stations and interchanges (nodes) on each line but does not attempt to reflect their true position nor the distance between them.

topsoil the upper, cultivated layer of soil, which may vary in depth from 8 to 45 cm. It contains organic matter – the decayed remains of vegetation, which plants need for active growth – along with a variety of soil organisms, including earthworms.

tor isolated mass of rock, usually granite, left upstanding on a hilltop after the surrounding rock has been broken down. Weathering takes place along the joints in the rock, reducing the outcrop into a mass of rounded blocks.

tornado extremely violent revolving storm with swirling, funnel-shaped clouds, caused by a rising column of warm air propelled by strong wind. A tornado can rise to a great height, but with a diameter of only a few hundred metres or less. Tornadoes move with wind speeds of 160–480 kph, destroying everything in their path. They are common in the central USA and Australia.

tourism visit to a place away from home that (unlike ◊recreation) involves at least an overnight stay. An area that attracts tourism can increase its wealth and job opportunities, although the jobs may be low paid and seasonal. Among the negative effects of tourism are traffic congestion and damage to the environment. In the UK, tourism generates £23 billion a year and accounts for 4% of ◊gross domestic product (GDP). It provides jobs for about 1.4 million people.

town settlement intermediate in size between a village and a city. Towns in the UK typically have a population of 4,000–90,000. They may be divided into distinct land-use sectors and eventually grow to form conurbations. A town may be dominated by one function; for example, an industrial town or a resort town.

trade wind ◊prevailing wind that blows towards the equator from the northeast and southeast.

traffic vehicles using public roads. In 1970 there were 100 million cars and lorries in use worldwide; in 1990 there were 550 million. One-fifth of the space in European and North American cities is taken up by cars. In 1989 UK road-traffic forecasts predicted that traffic demand would rise between 83% and 142% by the year 2025.

transform fault see ◊faulting.

transhumance seasonal movement by pastoral farmers of their livestock between areas of different climate. There are three main forms: in ***Alpine*** regions, such as Switzerland, cattle are moved to high-level pastures in summer and returned to milder valley pastures in winter; in ***Mediterranean*** lands, summer heat and drought make it necessary to move cattle to cooler mountain slopes; in ***W Africa***, the nomadic herders of the ◊Fulani peoples move cattle south in search of grass and water in the dry season and north in the wet season to avoid the ◊tsetse fly.

transition zone in a city, the area surrounding the central business district, characterized by ageing industry, derelict land, and low-cost housing; it is the edge of the ◊inner city.

transmigration relocation of large numbers of people away from overpopulated core regions (see ◊core and periphery) to less crowded areas. In Indonesia since 1986 there have been government-sponsored ◊resettlement schemes from such islands as Java, where population densities and unemployment are high, to the outer islands, especially Sumatra and Sulawesi, where land is less densely populated. Unfortunately, this has resulted in massive ◊deforestation of these areas.

transpiration in plants, the process by which water is given off as vapour from leaves. With ◊evaporation it forms the means by which water moves from the Earth's surface to the atmosphere (see ◊hydrological cycle).

Transpiration is most significant over forest areas, and the cutting down of large areas of forest (◊deforestation) can result in the local climate becoming less humid.

tremor minor ◊earthquake.

tributary river that joins a larger river; for example, the river Teme is a tributary of the river Severn.

tropical cyclone another term for ◊hurricane.

tropical monsoon see ◊monsoon.

tropics the area between the tropics of Cancer and Capricorn, defined by the parallels of latitude approximately 23°30' N and S of the equator. They are the limits of the area of Earth's surface in which the Sun can be directly overhead. The mean monthly temperature is over 20°C.

troposphere lower part of the Earth's ◊atmosphere extending about 10.5 km from the Earth's surface.

truncated spur blunt-ended ridge of rock jutting from the side of a glacial trough, or valley. As a glacier moves down a river valley it is unable to flow around the ◊interlocking spurs that project from either side, and so it erodes straight through them, shearing away their tips and forming truncated spurs.

tsetse fly blood-feeding insect that carries trypanosomiasis, or sleeping sickness. It is a serious pest in parts of W Africa and is partly responsible for the ◊transhumance (movement from pasture to pasture) of the ◊Fulani people. The disease can kill both animals and people. Moist swampy conditions favour the fly.

tsunami wave generated by an undersea ◊earthquake or volcanic eruption. In the open ocean it may take the form of several successive waves, in excess of a metre in height. In the coastal shallows tsunamis slow down and build up, producing towering waves that can sweep inland and cause great loss of life and property. In 1983 an earthquake in the Pacific caused tsunamis up to 3 m high, which killed over 100 people in Akita, N Japan.

tundra region of high latitude almost devoid of trees, resulting from the presence of ◊permafrost. The vegetation consists mostly of grasses, sedges, heather, mosses, and lichens. Tundra stretches in a continuous belt across N North America and Eurasia.

twilight zone another term for ◊transition zone.

typhoon violently revolving storm, a ◊hurricane in the W Pacific Ocean.

U

underemployment inefficient use of the available workforce. Underemployment exists when many people are performing tasks that could easily be done by fewer workers, or when a large percentage of the workforce is unemployed.

undernourishment condition that results from consuming too little food over a period of time. Like ***malnutrition*** – the result of a diet that is lacking in certain nutrients (such as protein or vitamins) – undernourishment is common in poor countries. Both lead to a reduction in mental and physical efficiency, a lowering of resistance to disease in general, and often to deficiency diseases such as beriberi or anaemia. In the developing world, lack of adequate food is a common cause of death.

urban decay the decline of the social, physical, and economic fabric of a city, usually located in the oldest part of the settlement – for example, London docklands – or the ◊inner city.

urban development corporation (UDC) in the UK, an organization set up by the central government to coordinate rapid improvements within depressed city areas. UDCs were first introduced 1981 in the London Docklands and Merseyside. Their aims are typically: (1) to improve the local environment, making it more attractive to business; (2) to give cash grants to firms setting up or expanding within the area; (3) to renovate and reuse buildings; (4) to offer advice and practical help to firms considering moving to the location.

urban fringe boundary area of a town or city, where new building is changing land use from rural to urban (see ◊urban sprawl). It is often a zone of planning conflict.

urbanization process whereby an increasing proportion of a region's population becomes concentrated in urban areas. It involves not only ◊migration from the countryside to cities, but also the growth of urban and suburban populations.

In many industrialized countries, urbanization was rapid in the 19th century; for example, in the UK, when employment became concentrated in towns and cities. More recently there has been the trend of ◊counterurbanization. In many developing countries, urbanization since the 1950s has been much faster than in the developed world. This has been caused by mass rural-to-urban migration, and high rates of ◊natural increase within the urban areas. The result is a number of cities in the South with populations exceeding a million (see ◊millionaire city).

urban land-use model in the social sciences, a simplified pattern of the land use (such as industry, housing, and commercial activity) that may be found in towns and cities. These models are based on an understanding of the way in which these areas have grown. There are three main ways of looking at urban land use: ◊concentric-ring theory, ◊sector theory, and ◊multiple-nuclei theory. Each results in different shapes of land-use areas. In practice, factors such as ◊topography, land fertility, and culture vary from one city to another and affect their final form.

urban renewal adaptation of existing buildings in towns and cities to meet changes in economic, social, and environmental requirements, rather than demolishing them.

Urban renewal has become an increasingly important element of inner-city policy since the early 1970s. A major objective is to preserve the historical and cultural character of a locality, but at the same time to improve the environment and meet new demands, such as rapidly increasing motor traffic. One option is ◊gentrification, raising an area's social and economic status.

urban sprawl outward spread of built-up areas caused by their expansion. This is the result of ◊urbanization. Unchecked urban sprawl may join cities into ◊conurbations; ◊green-belt policies are designed to prevent this.

Increased mobility and improved transport systems in the developed countries such as the UK have encouraged urban growth that is often linear in form (see ◊linear development).

U-shaped valley another term for a ◊glacial trough, a valley formed by a glacier.

V

veldt subtropical grassland in South Africa, equivalent to the ◊Pampas of South America.

velocity of flow the distance travelled by water per unit of time (usually, per second). The velocity of flow along a particular river channel is used to calculate the discharge of that channel (the volume of water passing a particular point in a given period of time), which in turn is used to assess the likelihood of flooding.

Velocity of flow is proportional to the gradient (slope) of the river channel – the steeper the gradient, the faster the velocity. However, average velocity tends to increase with distance downstream, even though the gradient gets less, because the river channel becomes smoother and is therefore more efficient (see ◊channel efficiency).

Velocity of flow may be measured by timing a simple ***float*** over a set distance (the accuracy of the measurement is increased if the process is repeated several times to produce an average figure). A more accurate (but more expensive) device is a ***flowmeter***, which consists in essence of a propeller that is spun by the flowing water. The number of spins is then related to the water's velocity.

vent hole in a ◊volcano through which magma, or molten rock, rises. If the magma is particularly thick it may clog up the vent, causing a tremendous buildup of pressure, which will be released as a very violent eruption.

village rural settlement intermediate in size between a ◊hamlet and a ◊town (population about 200–3,000). The term may also refer to the village-style parts of larger urban areas.

Although the primary function of a village is residential, it will normally contain a church, a pub, and other low-order services (see ◊hierarchy) such as a post office and general store.

volcanic rock another name for ◊extrusive rock.

volcano crack in the Earth's crust through which hot magma (molten rock) and gases well up. The magma becomes known as lava when it reaches the surface. A volcanic mountain, usually cone shaped with a crater on top, is formed around the opening, or vent, by the build-up of solidified lava and ashes (rock fragments). Most volcanoes arise on plate margins (see ◊plate tectonics), where the movements of plates generate magma or allow it to rise from the mantle beneath. However, a number are found far from plate-margin activity, on ◊hot spots where the Earth's crust is thin. There are about 600 active volcanoes on Earth. Some volcanoes may be inactive (dormant) for long periods.

volcano

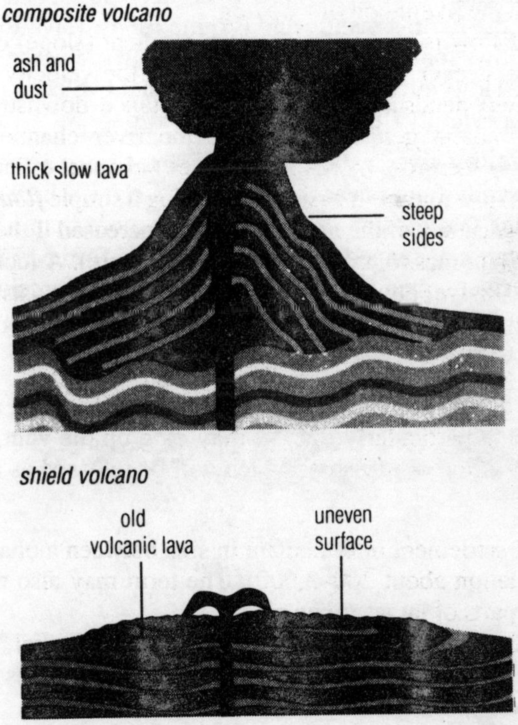

There are two main types of volcano.
Composite volcanoes, such as Stromboli and Vesuvius in Italy, are found at destructive plate margins (areas where plates are being pushed together), usually in association with island arcs and coastal mountain chains. The magma is mostly derived from plate material and is rich in silica. This makes it very stiff and it solidifies rapidly to form a high, steep-sided volcanic mountain. The magma often clogs the volcanic vent, causing violent eruptions as the blockage is blasted free, as in the eruption of Mount St Helens, USA, in 1980. The crater may collapse to form a ◊caldera.
Shield volcanoes, such as Mauna Loa in Hawaii, are found along the rift valleys and ocean ridges of constructive plate margins (areas where plates are moving apart), and also over hot spots. The magma is derived from the Earth's mantle and is quite runny. Lava formed from this magma flows for some distance over the surface before it sets and so forms broad low volcanoes. The lava of a shield volcano is not ejected violently but simply flows over the crater rim.

V-shaped valley river valley with a V-shaped cross-section. Such valleys are usually found near the source of a river, where the steeper gradient means that erosion cuts downwards more than it does sideways. However, a V-shaped valley may also be formed in the lower course of a river when its powers of downward erosion become renewed by a fall in sea level, a rise in land level, or the capture of another river (see ◊rejuvenation).

W

water cycle another name for the ◊hydrological cycle, by which water is circulated between the Earth's surface and its atmosphere.

waterfall cascade of water in a river or stream. It occurs when a river flows over a bed of rock that resists erosion; weaker rocks downstream are worn away, creating a steep, vertical drop and a plunge pool into which the water falls. As the river ages, continuing erosion causes the waterfall to retreat upstream forming a deep valley, or ◊gorge.

waterfall

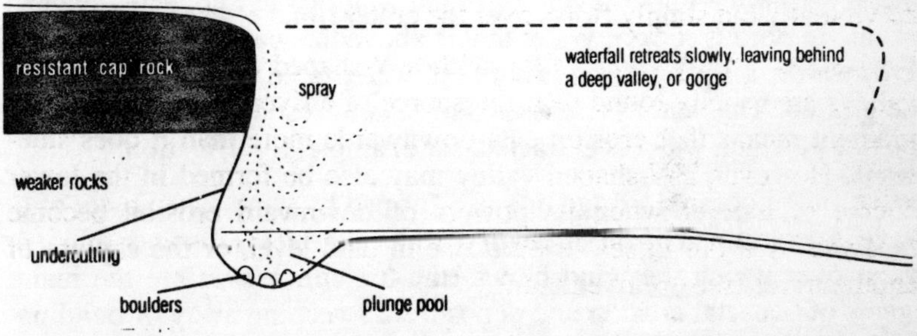

watershed the boundary between two ◊drainage basins, usually a ridge of high ground.

water supply distribution of water for domestic, municipal, or industrial consumption. Water supply in sparsely populated regions usually comes from underground water rising to the surface in natural springs, supplemented by pumps and wells. Urban sources are deep artesian wells, rivers, and reservoirs, usually formed from enlarged lakes or dammed and flooded valleys, from which water is conveyed by pipes, conduits, and aqueducts to filter beds. As water seeps through layers of shingle, gravel, and sand, harmful organisms are removed and the water is then distributed by pumping or gravitation through mains and pipes. Often other substances are added to the water, such as chlorine and fluorine; aluminium sulphate, a clarifying agent, is the most widely used chemical in water treatment. In towns, domestic and municipal (road washing, sewage) needs account for about 135 l per head each day. In coastal desert areas, such as the Arabian peninsula, desalination plants remove salt from sea water. The Earth's waters, both fresh and saline, have been polluted by industrial and domestic chemicals, some of which are toxic and others radioactive.

In 1989 the regional water authorities of England and Wales were privatized to form ten water and sewerage companies. Following concern that some of the companies were failing to meet EC drinking-water standards on nitrate and pesticide levels, the companies were served with enforcement notices by the government Drinking Water Inspectorate.

water table the upper level of ground water (water collected underground in porous rocks). Water that is above the water table will drain downwards; a spring forms where the water table cuts the surface of the ground. The water table rises and falls in response to rainfall and the rate at which water is extracted, for example, for irrigation.

wave in the oceans, a ridge or swell formed by wind. The power of a wave is determined by the strength of the wind and the distance of open water over which the wind blows (the ◊fetch). Waves are the main agents of ◊coastal erosion and deposition: sweeping away or building up beaches, creating ◊spits and ◊berms, and wearing down cliffs by their hydraulic action and by the corrasion of the sand and shingle that they carry.

As a wave approaches the shore it is forced to break as a result of friction with the sea bed. When it breaks on a beach, water and sediment are carried up the beach as ***swash***; the water then drains back as ***backwash***.

A ***constructive wave*** causes a net deposition of material on the shore because its swash is stronger than its backwash. Such waves tend be low and have crests that spill over gradually as they break. The backwash of a ***destructive wave*** is stronger than its swash, and therefore

wave

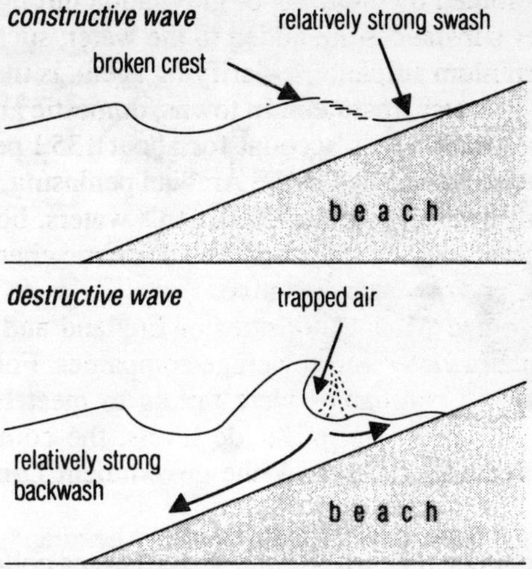

causes a net removal of material from the shore. Destructive waves are usually tall and have peaked crests that plunge downwards as they break, trapping air as they do so.

If waves strike a beach at an angle the beach material will be gradually moved along the shore (◊longshore drift), causing a deposition of material in some areas and erosion in others.

wave-cut platform gently sloping rock surface found at the foot of a coastal cliff. Covered by water at high tide but exposed at low tide, it represents the last remnant of an eroded headland (see ◊coastal erosion).

wave power power obtained by harnessing the energy of water waves. A number of wave-power devices have been advanced (such the duck – a floating boom whose segments bob up and down with the waves, driving a generator), but few have yet proved economical. A major breakthrough will be required if wave power is ever to contribute significantly to the world's energy needs.

wave refraction the distortion of waves as they reach the coast, due to variations in the depth of the water. It is particularly evident where there are headlands and bays.

wave refraction

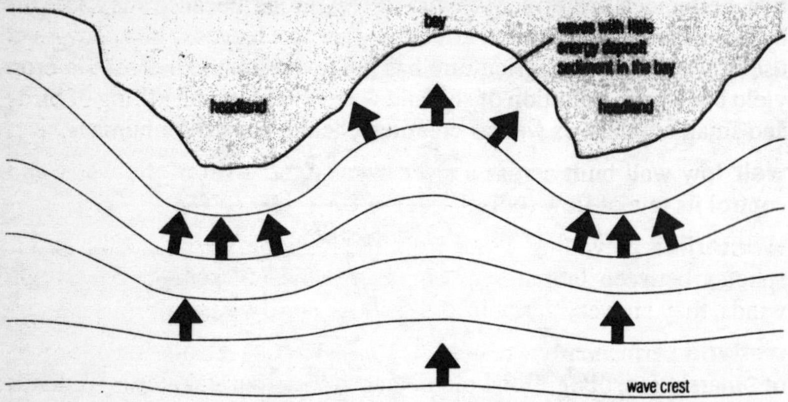

The bending of a wave crest as it approaches a headland concentrates the energy of the wave in the direction of that headland, and increases its power of erosion. By contrast, the bending that a wave crest experiences when it moves into a bay causes its energy to be dissipated away from the direction of the shore. As a result the wave loses its erosive power and becomes more likely to deposit sediment on the shore.

weather the day-to-day variation of atmospheric conditions at any one place. Such conditions include humidity, precipitation (rain, snow, and so on), temperature, cloud cover, visibility, and wind.

A region's ◊climate is derived from the average weather conditions over a long period of time.

weathering process by which exposed rocks are broken down on the spot by the action of rain, frost, wind, and other elements of the weather. It differs from ◊erosion in that no movement or transportion of the broken-down material takes place. Three types of weathering are recognized: ◊physical weathering, ◊chemical weathering, and ◊biological weathering. They usually occur together.

weather vane instrument that shows the direction of wind. Wind direction is always given as the direction from which the wind has come – for example, a northerly wind comes from the north.

weedkiller or ***herbicide*** chemical that kills some or all plants. Selective herbicides are effective with cereal crops because they kill all broad-leaved plants without affecting grasslike leaves. The widespread use of weedkillers in agriculture has led to a dramatic increase in crop yield but also to pollution of soil and water supplies and killing of birds and small animals, as well as creating a health hazard for humans.

weir low wall built across a river to raise the level of the water and control its rate of flow (◊discharge).

Westerlies prevailing winds from the west that occur in both hemispheres between latitudes of about 35° and 60°. Unlike the ◊trade winds, they are very variable and produce stormy weather.

wetland permanently wet land area or habitat. Wetlands include areas of ◊marsh, fen, ◊bog, flood plain, and shallow coastal areas. Wetlands are extremely fertile. They provide warm, sheltered waters for fisheries, lush vegetation for grazing livestock, and an abundance of wildlife.

Many wetlands are threatened by development – for example, the Camargue in France, which is being developed for rice cultivation.

wetted perimeter length of that part of a river's cross-section that is in contact with the water. The wetted perimeter is used to calculate a river's hydraulic radius, a measure of its ◊channel efficiency.

wetted perimeter

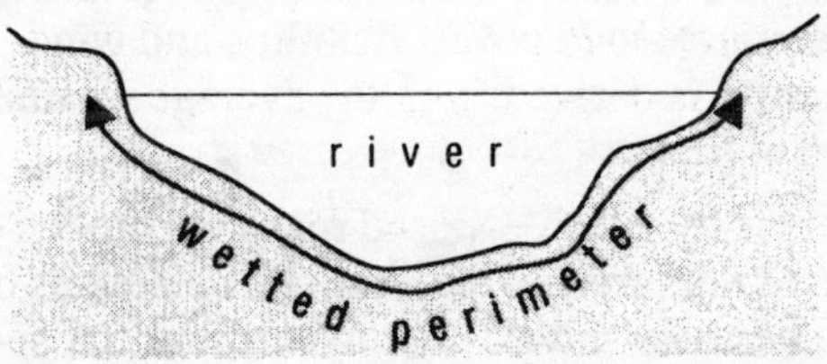

wheat cereal plant widely cultivated in temperate climates; it is the chief cereal used in breadmaking. Wheat is killed by frost, and damp renders the grain soft, so warm, dry regions produce the most valuable grain.

The main wheat-producing areas of the world are the Ukraine, the prairie states of the USA, the Punjab in India, the prairie provinces of

Canada, parts of France, Poland, S Germany, Italy, Argentina, and SE Australia.

whiteout condition of low visibility caused by heavy snowfall and strong winds. The uniform whiteness of the ground and air causes disorientation.

wind the horizontal movement of air across the surface of the Earth. Wind is caused by the movement of air from areas of high atmospheric pressure (◊anticyclones) to areas of low pressure (◊depressions). In the northern hemisphere, air moves out of high-pressure areas in a clockwise direction, and into low-pressure areas in an anticlockwise direction. Very strong winds, such as those associated with ◊hurricanes, can cause a great deal of damage, killing people and blowing down buildings and crops.

On a global scale several major wind patterns can be identified, including the trade winds; however, these are modified locally by land and water. Famous winds include the chinook of the North American Rockies and the föhn of Europe's Alpine valleys: both are warm winds.

Wind speed is measured using an ◊anemometer. It can also be measured by studying its effects on, for example, trees by using the ◊Beau-

wind

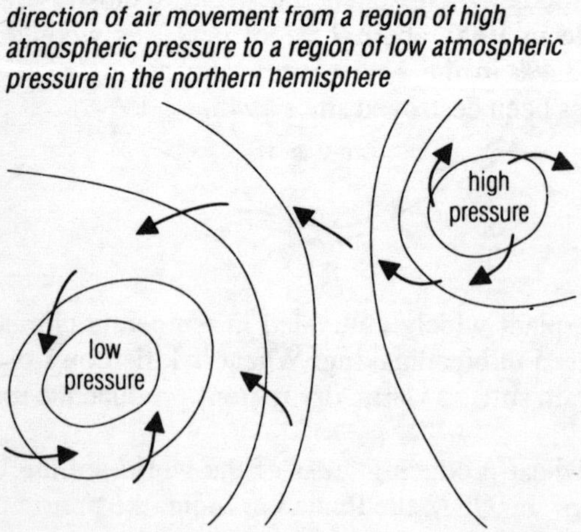

fort scale. Wind direction is measured using a ◊weather vane. The prevailing (most common) wind direction in the UK is the southwest.

wind energy energy derived from the wind. It is harnessed by sailing ships and windmills, both of which are ancient inventions. Wind turbines are aerodynamically advanced windmills that drive electricity generators when their blades are spun by the wind. Wind energy is a renewable resource that produces no direct pollution of the air; it is therefore beginning to be used to produce electricity on a large scale.

wind farm large area of land covered by windmills or wind turbines, used for generating electrical power. A wind farm at Altamont Pass, California, USA, consists of 300 wind turbines. To produce 1,200 megawatts of electricity (an output comparable with that of a nuclear power station), a wind farm would need to occupy around 370 sq km.

In the UK the first commercial windfarm, near Camelford, Cornwall, began to generate electricity in 1992.

woodland area in which trees grow more or less thickly; generally smaller than a forest. Temperate climates, with four distinct seasons a year, tend to support a mixed woodland habitat, with some conifers but mostly broad-leaved and deciduous trees, shedding their leaves in autumn and regrowing them in spring. In the Mediterranean region and parts of the southern hemisphere, the trees are mostly evergreen.

In England in 1900, about 2.5% of land was woodland, compared with about 3.4% in the 11th century. An estimated 33% of ancient woodland has been destroyed since 1945.

Y

young fold mountain one of a range of mountains formed by the crumpling of the Earth's crust at a destructive margin (where two plates

young fold mountain

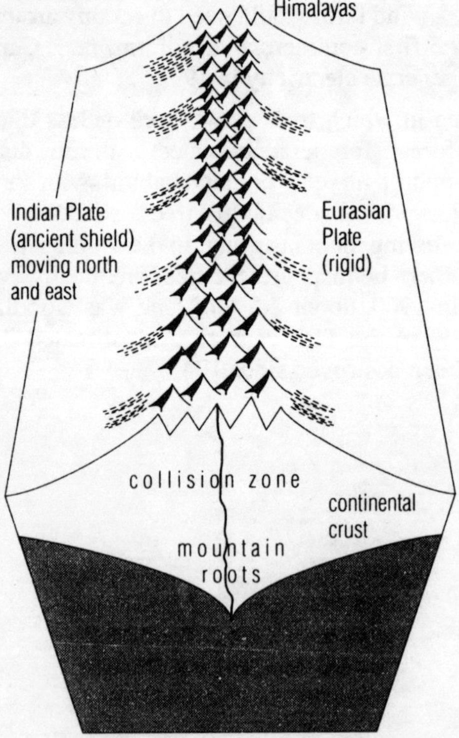

collide). For example, the Himalayas of central Asia, which have formed over the last 50 million years at the margin between the Indian and Eurasian plates (see ◊plate tectonics). Other ranges of young fold mountains are the European Alps, the Andes of South America, and the Rockies of North America.

Fold mountains are formed when the continental crust of the two colliding plates is relatively buoyant; neither plate will therefore subduct, or descend, beneath the other. The resulting pressure between the plates causes the crustal material to become folded and uplifted.

Young fold mountains are very active areas: earthquakes and landslides are common, and in the Andes range there are several active volcanoes, such as Nevado del Ruiz in Colombia.